Nancy McMahon

Water

The Element of Life

Theodor Schwenk on the Mississippi, 1982

Water

The Element of Life

Essays by

Theodor Schwenk and Wolfram Schwenk

Translated by Marjorie Spock

Anthroposophic Press

Published in the United States by Anthroposophic Press, Bell's Pond, Star Route, Hudson, New York 12534.

Library of Congress Cataloging-in-Publication Data

Schwenk, Theodor.
Water: the element of life: essays / by
Theodor Schwenk and Wolfram Schwenk; translated by Marjorie Spock.
p. cm.
Essays originally published in various journals in German. Includes bibliographical references.
ISBN 0-88010-277-2
1. Water—Religious aspects. 2. Anthroposophy. I. Schwenk, Wolfram. II. Title.
BP596.W38s38 1989
299'.935—dc20 89–36426

Printed in the United States of America

Contents

Acknowledgments

"What Is Living Water?" first appeared in German under the title *"Zum Begriff des Lebendigen Wassers,"* copyright 1967 by Institute for Flow Sciences, Herrischried. It was previously published in English in *The Golden Blade* (London: Hawthorn Press, 1969). It is reprinted here with kind permission of the publishers.

"Water Consciousness" was originally published in German as *"Das Wasser, Element der Schwelle im modernen Bewusstsein,"* copyright 1972 by Institut for Flow Sciences, Herrischried. It was previously published in English in the *Journal for Anthroposophy* 18, 1973. It is reprinted here with kind permission of the publishers.

"The Spirit in Water and in Man" was published in German as *"Von der hohen Schule des Wassers,"* copyright 1972 by Intitute for Flow Sciences, Herrischried. It was previously published in English in the *Journal for Anthroposophy* 26, 1977. It is reprinted here with kind permission of the publishers.

"Water: Destiny of the Human Race" appeared in German under the title *"Das Wasser, Schicksal der Menschheit,"* copyright 1976 by Institute for Flow Sciences, Herrischried. It appeared previously in English translation in the *Journal for Anthroposophy* 30, 1979 (fall). It is reprinted here with kind permission of the publishers.

"The Warmth Organism of the Earth" first appeared in German under the title *"Vom Wärmeorganismus der Erde,"* copyright 1974 by Weleda, Schwäbisch Gmünd. It was previously published in English in the *Journal for Anthroposophy* 23, 1976, and it is reprinted here with kind permission of the publishers.

"Keeping the Earth the Place of Life" was published

in German under the title *"Wie kann die Erde Schauplatz des Lebens bleiben?"* copyright 1976 by Institute for Flow Sciences, Herrischried. It was previously published in English translation in the *Journal for Anthroposophy* 40/41, 1985 (spring). It is reprinted here with kind permission of the publishers.

"The Human Role in the Earth's New Connection with the Cosmos" was originally published in German as *"Die neue Verbindung der Erde mit dem Kosmos durch den Menschen,"* copyright 1977 by Institute for Flow Sciences, Herrischried. It is published here for the first time in English.

"Motion Research: Its Course and Aims over Twenty Years" was published in German under the title *"Zwanzig Jahre Verein für Bewegungsforschung, seine Ziele und Wege,"* copyright 1981 by Institute for Flow Sciences, Herrischried. It is here published for the first time in English translation.

The nine articles listed above are collected in the first volume of the German series *Sensibles Wasser,* (Herrischried: Association for Motion Research, Institute for Flow Sciences, 1985). The Institute kindly gave permission to publish these articles in this collection.

"Water as the Element of Life" was published in German under the title *"Wassernot und Wasserrettung: Unser Wasser als Lebenselement,"* copyright 1982 by Verein für ein erweitertes Heilwesen e.V., Bad Liebenzell. It is here published for the first time in English.

"Water Sustains All" was originally published in German as *"Alles wird durch das Wasser erhalten: Ein Beitrag zum Verständnis des Lebenselementes Wasser,"* copyright 1977 by Institut for Flow Sciences, Herrischried. A previous translation of this essay was published under the title "All Things Are Maintained by Water" in *Journal of Anthroposophic Medicine* 1, Stuttgart 1981.

"Testing for Water Quality: The Drop-Picture Method" appeared in German under the title "*Qualitätsprüfung des Wassers mit der Tropfenbildmethode*" in *Das Seminar: Zeitschrift für Fachfortbildung und Praxisinformation* copyright 1980 by Hessisches Winterseminar, Wiesbaden. It is here published for the first time in English.

"Studying the Behavior of Water" was published in German under the title "*Zur Verhaltensforschung des Wassers*" in *Der Aufbau: Fachschrift für Planen, Bauen, Wohnen und Umweltschutz* 3–5/ 1979, copyright 1979 by *Der Aufbau*, Wien, Austria.

"Water as a Nutrient" was published in German as "*Lebensmittel Wasser: Die Frage nach seinen Qualitäten*" in *Die Drei* 7/8 (July/August) 1982, copyright 1982 by *Die Drei*.

All drop-picture photographs are property of the Institute for Flow Research, Herrischried, and are reproduced here with the permission of the Institute. Figures 1–4 in "Water as the Element of Life" are reproduced from Theodor Schwenk's *Sensitive Chaos* with permission of Verlag Freies Geistesleben, Stuttgart.

Foreword

Theodor Schwenk opened up new perspectives on water and its formative capacity in movement with his book *Sensitive Chaos,* which reached a large worldwide readership. Many English readers expressed a wish to have translations of his later lectures and of other publications that set forth the findings from his investigations of water.

That this wish could be met is due in large part to Marjorie Spock and her translations of lectures and articles of both Thoedor Schwenk and his son, Wolfram Schwenk. With sensitivity for their styles and their natural- and spiritual-scientific concepts, she translated their works, served as an intermediary between the authors and the publisher, and found interested friends to finance the publication project. She responded to every wish and suggestion of the authors with understanding. Working with her has been a wonderful experience and I want to express my special gratitude to her here.

Theodor Schwenk was born in 1910 in Schwäbisch-Gmünd, West Germany. He studied engineering and earned his diploma in the field of hydrotechnology, working for several years on research and development in that industry. In 1937, he was appointed assistant to the rheologist Ludwig Prandtl in Göttingen. He became interested in Rudolf Steiner's anthroposophy early in life and looked for opportunities to serve it professionally. In 1946 he was employed by Weleda, the manufacturer of anthroposophical pharmaceuticals in Schwäbisch Gmünd, where he engaged in investigations leading to a better understanding of the potentizing of medical substances. His book *The Basis of Potentization Research*, published in 1954, which detailed his findings, has recently appeared in English translation.* During that period Theodor Schwenk worked over the rheological experiences and insights he had gained and wrote them down in *Sensitive Chaos*, which was first published in 1962.

In 1959 Theodor Schwenk started laying the foundation for the *Institut für Strömungswissenschaften* (Institute for Flow Sciences), located at Herrischried in the southern part of the Black Forest, and it opened in 1961. It is a small private undertaking under the aegis of the *Verein für Bewegungsforschung* (Association for Motion Research), and is supported by donations. Its history and an account of the early goals it set itself are described in the eighth chapter of this book, "Motion Research".

Theodor Schwenk delivered the lectures that appear here in English translation to friends and supporters of the Institute. They give a clear picture of his purpose and the route he followed to a perception of the earth as a vast living organism, an organism only to be understood in a discoverable relationship to its cosmic environment. He says in one of the lectures that "the statements made in these lectures are intended to point again and again to organic relationships of a large order, and to convey information from which responsible practical undertakings can be developed."

The integrating and interpreting of phenomena from the standpoint of the anthroposophically oriented holistic thinking that has been particularly prized in *Sensitive*

The Basis of Potentization Research, (Spring Valley, N.Y.: Mercury Press, 1988).

Chaos comes to expression in these lectures in a clearly more personal way. It is a thinking capable of providing a variety of incentives and pointers for an understanding of water's essential being. And the lectures may also be looked upon as a living documentation of Theodor Schwenk's wrestling with the sleepiness to which people of that period had succumbed in their way of dealing with water.

At the Institute in Herrischried Theodor Schwenk worked out the drop-picture method for determining water quality, especially in the case of drinking water, and published the results of his research in the book *Bewegungsformen des Wassers*. The drop-picture method opened up a field of science so new and important that it remains to this day the area upon which the *Institut für Strömungswissenschaften* focuses its chief interest. Theodor Schwenk himself retired as its director in 1976, handing over his responsibilities to his son Wolfram Schwenk, who had worked as a water biologist at the Institute since 1972. From 1977 to 1984, Theodor Schwenk devoted himself to problems of the regeneration of water by the introduction of processes of motion. This work was carried on in a laboratory at Neustadt an der Weinstrasse. In 1985 he retired to Stuttgart, where he died on September 29, 1986.

The work of the Institute, which employs three scientists, focuses on efforts to determine what makes for living qualities in good drinking water and to base a renewal of awareness of water as a life-giving element on such information. This field of investigation stands in a very real sense in the mainstream of contemporary science with its research into the bases and conditions of formative processes in fluid systems. The Institute's scientific research is oriented to Rudolf Steiner's anthroposophical spiritual science.

The articles written by Wolfram Schwenk provide examples of this effort. Theodor Schwenk's lecture entitled "The Warmth Organism of the Earth," delivered in 1974 to physicians at a conference at Weleda, Inc., is so closely related to the Herrischried lectures that it found inclusion here. Repetitions that crop up here and there may serve as valuable means for deepening readers' awareness of certain aspects treated in others of the lectures. The article translated as "Water as the Element of Life" was written by Theodor Schwenk for the popular scientific magazine *Soziale Hygiene,* issued by the *Verein für ein erweitertes Heilwesen,* whose purpose it is make anthroposophical viewpoints available to a wider readership.

The articles included in this collection were published individually by different presses. They appear here for the first time in collected form. Warm thanks go to their publishers and editors for permission to reprint them. They are named in the Acknowledgments. Thanks go also to the Anthroposophic Press for publishing this American edition.

April 1989 WOLFRAM SCHWENK

What is "Living Water"?

Theodor Schwenk

Those who keep up with developments in any phase of today's water problem constantly come across references to "living water" and "dead water." For some, these are perfectly familiar terms and concepts they use every day in their professional practice. Others reject them on the ground that since water possesses none of the characteristics of living organisms it cannot be spoken of as either alive or dead.

Big-city dwellers who go into mountain country and drink from a ceaselessly bubbling spring in some mountain meadow know firsthand what living water is. Precisely because they come from a big city they really can distinguish between dead and living water. They know, of course, that city water has to meet strict sanitary standards. But they have had personal experience of the vast difference between that water and what they drink from mountain springs. And though what comes out of city faucets can be counted on not to contain coliform bacteria or other toxic organisms and is sure to be free of nitrates and phosphates, can these criteria suffice when it comes to characterizing what they know firsthand as living water?

In dealing with modern water problems, the inevitable question arises of how to describe and verbalize the

---

Digest of a lecture given at Herrischried, West Germany, on September 3, 1967, for friends and supporters of the *Institut für Strömungswissenschaften*, the Institute for Flow Sciences.

concept of living and dead water . . . despite immediate experience and the familiarity of those terms.

We will try to work out such a concept in the following pages.

Let us begin by recalling some unspoiled brook that we may have seen making its way through dark forest depths, now burbling over pebbly stretches, now pent up in quiet pools. It sparkles in the changing play of light that breaks through the screen of foliage above; it leaps ahead in rippling wavelets, alternating between soft murmuring and silvery tinkling. It takes a meandering course among the trees, twisting this way and that as though to make its lively game last longer. Surely this water cannot be called anything but living!

If we pick up a stone along the bank, its wet underside and the hole it came out of will be found all a-wriggle with tiny living creatures. Should water that contains such life be designated "living?" We certainly do not want the water that comes out of our faucets full of it, even though its presence is considered by science to be one of the best indicators of a water's excellence, and scientific classifications are actually based on the presence or absence of that life.

Nevertheless, none of it is welcome in our drinking water, whether it comes out of faucets or from mountain springs. The concept "living water" must be built on some other sort of criteria.

What is it that attracts us so powerfully to living water? Do we perhaps feel that water flowing out of faucets in dark city flats is dead and almost unreal because of imponderable elements in the environment? Whereas water that constantly pours out of a spring in the middle of a flower-filled mountain meadow, sparkling in the pure, cold air and in light reflected from shimmering snow peaks round about, perfectly satisfies our concept of what living water is.

Our age feels the need to form concepts of even the most livingly experienced realities. And so too in the case of living water, which we want to *grasp* in the double sense both of understanding what it is and how to have it and hold on to it.

But if we take this grasping literally and try to take hold of water, it slips through our hands and flows away unless we put it in some sort of container. If the container is solid we can "catch" water in it; then it quiets down immediately. But this means that it is already well on the way to being dead, to losing the quality that originally made it living water. And the concepts we form of it suffer the same fate: the moment one tries to catch the real nature of water in a hard and fast definition, the reality of it is no longer there. To be adequate, concepts of water must retain something of water's liveliness and movement, of the way it lends itself to constant change. They must be capable of metamorphosis, shaped in harmony with, and able to express, water's functions.

If we ask again at this point, "What, specifically, is living water?" we see that another question has to be answered first, namely, "What is life?"

Life manifests itself in quite specific, tangible attributes in every living creature: in growth, reproduction, metamorphosis (i.e. changes of form that take place in organisms); in metabolic functions, which include the digesting of food and excreting of wastes; in the regulation of chemical, warmth and other processes; in tides of burgeoning and fading that rise and fall in definite time patterns, always subject to life's characteristic rhythms.

Are these attributes of life to be found in water? Does

water grow? Can it reproduce? We certainly know water to be the very prototype of everything formless and fluidic; how, then, could it possibly be said to undergo metamorphosis, to change its form, as living organisms do? Is it subject to digestive and excretory processes? Does it possess definite organs that perform typical organ functions? Can it adapt itself to fixed determining factors? Does water evolve into an organism and then die? Are autonomous rhythms, such as the heartbeat of an animal, to be found in water?

The answer to all these questions is of course obvious: water possesses none of these characteristics of living organisms. Whatever it turns out to be, we feel sure that it must rank lower than a living creature.

But if that is so, must our search for an adequate definition of living water not end up in a blind alley?

It would seem so. Yet what of the experience of people who work professionally with water and daily apply the concept "dead" and "living" water in their practice? Perhaps we had better wait to draw conclusions until we have looked at the problem from another side.

Everyone knows that water has a close affinity with all forms of life. So we may ask what the connection is—whether manifestations of life can occur in the absence of water. Is growth possible without it? Propagation? Metamorphosis? Are digestion and excretion conceivable where water is lacking? Is it not essential to conversion processes and organ functions? Is it not the great mediator and regulator of chemical, warmth and other processes, both in living organisms and outside them? Can a living organism even come into being without water's help? And, in the last analysis, could rhythm exist in organisms if there were no fluids?

Here again, to put the question is to find that the answer is self-evident: none of the above life-characteristics would be present without water. Life depends on water for its very life! Again we are confronted by a riddle: Water does not manifest a single life-characteristic. Yet where water is lacking, there can be no life. All the factors noted depend upon water.

At this point the ground beneath our feet begins to seem somewhat less than solid, as can often happen with true riddles. But isn't it perfectly natural to feel at sea when pursuing the theme of water? Perhaps this state of affairs may even prove an asset and set us on the right course for our further inquiry.

Why, then, does water, which has no life-characteristics of its own, form the very basis of life in all life's various manifestations? Because water embraces everything, is in and all through everything; because it rises above the distinctions between plants and animals and human beings; because it is a universal element shared by all; itself undetermined, yet determining; because, like the primal mother it is, it supplies the stuff of life to everything living.

And what makes water capable of all these feats?

- Renouncing any form of its own, it becomes the creative matrix for form in everything else.
- Renouncing any life of its own, it becomes the primal substance of all life.
- Renouncing material fixity, it becomes the implementer of material change.
- Renouncing any rhythm of its own, it becomes the progenitor of rhythm elsewhere.

Is it any wonder, then, that in all highly developed cultures water has always been held sacred as the magically transforming substance, as the very "water of life?"

Now that we have considered the plant, animal, and human kingdoms as specimens of living organisms, let us try to come a step closer to the answer to our question by bringing that all-inclusive living being, the earth-organism, into the picture as the living whole it is.

There exist comprehensive discussions of the earth as a living entity in which it is shown that such a concept is no mere theory, but a reality quite susceptible of human experience. These studies range from works by the great astronomer Johannes Kepler to contemporary presentations by Guenther Wachsmuth, Walther Cloos and others.[1]

Contemplating this vast living organism earth, one's attention is drawn again and again to the layer-structure of its great enveloping mantles and to the rhythms that play in and through them. A glance at the surface configuration of the earth reminds us, for example, that 70% of it is covered by water. This watery surface, in its immense extension, provides a plane of contact with the atmosphere. Here an exchange sets in between the elements of air and water that moves in both directions, up and down. Thus, water is absorbed into the atmosphere, where it works as the great regulator in matters of climate and in meteorological processes and their rhythms. Meteorologists, whose daily observation of the weather and its changes has led to the development of a special sense, shared by all outdoorsmen, for what goes on in the atmosphere, often find that they have to speak of processes there as of something living. August Schmauss, for example, talks of "biological concepts in meteorology," of an "orchestral score" of atmospheric happenings, with "entrances" in the time-pattern of the year's unrolling seasons.[2] Paul Raethjen says in his treatise on the dynamics of cyclones that "the atmosphere behaves like a living creature," and elsewhere in the same work we read:

> For one thing, cyclones have a metabolic process without which they could not exist: they constantly draw new masses of air into their vortices and excrete other masses in their outward-spirallings. Then too, . . . they have a typical life history with characteristic beginning, developing, and aging phases. They reproduce themselves, not in a wave-like spreading out in space, but like a living creature, in the sense that a young "frontal cyclone" is born out of the womb of an adult "central cyclone."[3]

We know that cyclones (low-pressure areas) have to do with water. A "low" and rain belong in the same concept.

Thus, the life of the earth-organism as a whole is just as closely bound up with water as the life of any of the creatures on it. Rhythmic processes are present in the cosmos that play into the various atmospheric strata, giving rise to the rhythms found wherever water is. Rudolf Steiner, Guenther Wachsmuth, Ernst Marti, George Adams, Hermann Poppelbaum, Lily Kolisko and others have acquainted us with these "formative forces" raying in from the cosmos; they have taught us to distinguish between them and to recognize how they build and shape all earth's living organisms.

Everything in nature forms one indivisible fabric woven of living interchange. An all-encompassing world of life comes into being from the interplay of cosmic peripheral forces, meteorological forces, forces of the elements, the earth, and all its living organisms.

A great deal of evidence has already been uncovered showing that the earth, and the various forms of life present on it, function in harmonious accord with universal processes, and every year more such evidence turns up. Almost every rhythm, from moon rhythms reflected in the hydrosphere and planetary rhythms known to meteorology right down to the numberless physiological rhythms found in every kind of living organism, is based on water's mediation. For example, woodcutters in the forests of Brazil still set the price of the wood they fell by the date of its cutting, that is, by the moon phase, because its water content (and thus its keeping quality) depends on these cosmic influences. The patterns of movement planets weave in space are also reflected in the structure of the various plant families; thus, for example, the Venus pattern appears in the regular pentagram common to all rose plants. If it were not for the mediating role water plays, these formative forces could not work their way into terrestrial manifestation. In the tides, the seas are caught up in the swing of cosmic rhythms that they then hand on to the earth and its creatures. All movement in water is affected by cosmic formative forces and serves the function of transmitting them.[4]

Thus, water occupies a median position between earth and the universe, and is the port of entry through which cosmic-peripheral forces pass into the earth realm.

May we not call water "nature's central organ," its "heart," the pulsing, oscillating drop that lets the whole cosmos pass through it? Its functions make it indeed the primal organ; it transmits cosmic forces and activities into the earth just as the heart mediates between processes of the upper and lower parts of the organism and—functionally speaking—embraces all the other organs.

Water does not grow because it is itself growth,

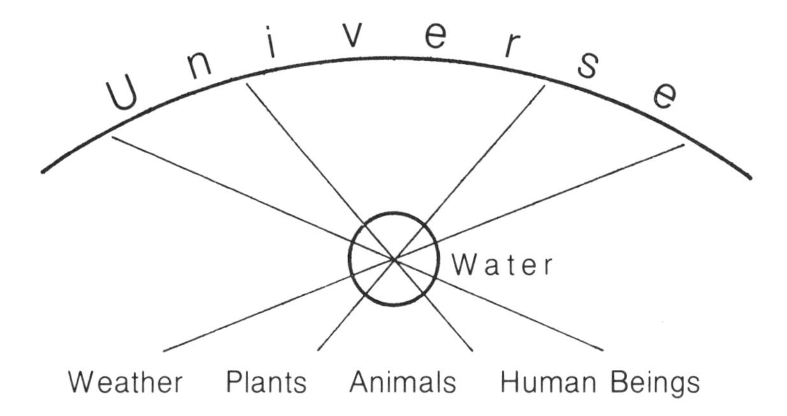

growth-as-function, growth uncommitted to any particular growth pattern. It is not subject to metamorphosis because it is itself the universal element of change that runs through every possible form without becoming fixed in any. While not itself subject to regulation, it serves as nature's regulator in numberless instances. Though it possesses no organs of its own, it is itself the primal organ jointly shared by everything that lives—an organ that remains at the functional level rather than presses on to the physical-organic stage. None of life's attributes come to outward expression in water, but they are all functionally present in it in the form of possibility, capacity, and action.

What wisdom we find built into man and nature!—like the wisdom that controls warmth in the bloodstream and keeps the normal level at precisely 98.6 degrees. Even a slight deviation from that temperature spells illness. How wisely it is arranged that water has the least specific warmth at exactly 98.6 degrees, and therefore absorbs warmth most rapidly at that temperature!

But water possesses many qualities beyond those

singled out above for mention, and these enable it to be the matrix of all living things. What a composition of wisdom every living organism is, each single facet of it revealing new marvels! And water is wisdom's very element, the focus in which wisdom is concentrated and out of which its activity flows into every least and greatest living thing. Indeed, it is because life is wisdom and water wisdom's element that there can be such a thing as the water of life.

Now that we have come to know water as the functional compendium of all outer forms of life, let us look from this new angle at some of the life-attributes discussed above.

Take nutrition, for example. Not only is the movement of foodstuffs through the organism unthinkable without the help of water but over 99% of all chemical and other changes depend on water. Drinking water of a quality that meets health requirements holds the balance between alkalinity and acidity.

Water, of course, occurs in greater volume in the ocean than anywhere else in nature. The composition of seawater is almost identical with that of blood, the only difference being that seawater contains magnesium where human blood contains iron—a fact that must be looked at in relation to the needs of air-breathing organisms.

What can be observed as the growth-process in trees, for example, is present in water as *pure function*. Moving water consists of layers that continually flow past one another at varying speeds. We find the same thing happening in the tree's cambium layer, where growth takes place. There too, newly formed cells slip past the old ones in an outward flow that widens the cambium ring without splitting it.

The vortex is the functional pattern of all the metamorphoses and shapings found in water. When layers flowing at different speeds pass one another, a process of deflection and rolling-up takes place, and this leads to the formation of a vortex. We find the same thing going on wherever growth-speeds vary in an organism. An example is the way a stem or leaf bends and starts to roll up when one side grows faster than the layer next to it.

Organs are also built on the model of the vortex-forming process, as may be noted in the way they always create a boundary between an inner and an outer space.

Reproduction, in the sense of repetition, is everywhere to be found in water, again of course at the functional level. Trains of vortices are an example. Here we see a process similar to the spawning of daughter-cyclones in a water-saturated atmosphere as described by Paul Raethjen.

In the case of rhythm, water must be called its very element. The word "rhythm" is derived from the Greek verb "to flow." And water does indeed flow rhythmically. This can be seen in the rhythmic pattern of waves and meandering watercourses, just as it can be heard in the audible rhythms of brooks and oceans. Numerical relationships found in physiological rhythms appear again in phenomena of air and water. Sound, for example, travels four times as fast through seawater as it does through air. This same proportion is found again in the rhythmic processes of man's fluid and airy organisms in the 4:1 pulse-*v.*-breathing rate.

Such facts make it evident how profoundly related water is to all life-functions, and most especially to those in the human organism.

Many of water's qualities also show a distinct relationship to soul and spiritual attributes in human beings.

Not only does water serve the cause of bodily life, but also that of the inner man in that, for example, the brain floats in water and is thus relieved of the pull of gravity. Physiologically speaking, this is what makes thinking possible.

Certain descriptions of water can equally well be applied to soul capacities; we speak of the power to cleanse, to purify, to heal, and both water and the soul are referred to as "in balance," "clear," "reflective," "refreshing." And just as healthy souls refrain from going to extremes, so water holds a balance between extremes of heavy and light with its buoyancy, between alkaline and acid, between warmth and cold, and—with its rainbow colors—between light and dark. And because it is so balanced an element it has the capacity to regulate, to heal, to create a mobile, living equilibrium. But when water comes into a resting state of balance, it stagnates, loses its "life," is as though paralyzed.

Since water is universal and a stranger to one-sidedness, it is the means whereby the full range of life is made possible. Hence, it is truly the "water of life." The concept "water" might best be expressed as "universality through renunciation," "wisdom's element." Because of the wisdom that makes it what it is, it is fitted to be the carrier of cosmic forces.

What, then, is living water?

It is water that contains not only the cosmic elements radiating life into the earth sphere, but that also has an inherent relationship to man as body, soul, and spirit.

Today this may sound like anthropomorphism. But there is a practical aspect to the statement. As we contemplate the civilization of our time, we can be aware of its water problems, but also of the direction we must take to solve them. If water is deprived of its universal nature by becoming tainted with substances like salts and detergents, it cannot maintain its universal function and continue to be the transmitter of nature's living wisdom. Wholesome water is water that maintains itself in active balance. But how far the sewage-laden rivers of today are from such a definition! They can no longer be recipients of cosmic forces. In many cases, indeed, the opposite is true: they can only be called "chemical infernos, veritable hells."

If we are to solve the water problems facing us today—and this includes such technical measures as must be taken—there is no other path open to us than to rediscover water's cosmic aspect. And this calls for nothing less than a new sense for what life is. From now on, everything depends on our developing what one professional in the field of water sanitation recently described as "water-consciousness." But that means recognizing that water is the carrier of cosmic energies and understanding how it can be rehabilitated to become once again their carrier.

Does this provide us with a concept of what living water is? We have come to know it as the bearer of wisdom, as the instrument of cosmic forces and orderings, in short, of life. We have found it to be living nature's primary organ, through which everything alive must pass.

While the work of our Institute takes its motivation from the restorative impulse vital to the solution of our time's great problems, it seeks that solution through grasping the true nature of water, seeks it, one might say, in a supersensible water-consciousness intent on taking the spirit of nature as the guide of its science and technology.

Water Consciousness

Theodor Schwenk

If we look into what has been going on in the worldwide water situation since our last meeting here two years ago, we will find that it has been steadily worsening. Efforts to clean up water pollution have simply not sufficed; the race is being won by pollution. The waters of the earth, the healing life-streams of our planet, have fallen ill and are presently engaged in a battle for their very lives.

Our earlier inquiry into what living water is ended with the report that foremost water experts are now stressing the fact that the great water problems confronting us are not just external ones, capable of solution by technical and economic means, but rather problems that can only be solved if every individual develops a new awareness and if the whole of humanity realizes that water is the element of life itself.

Such a challenge cannot be taken lightly; we must try to grasp it in its full significance. We will make that attempt in the following commentary.

One of the first questions to be asked is, of course, how a new "water consciousness" can be created. To answer this concretely we need to ask a further question: What sort of "water consciousness" do we have at

---

Digest of a lecture delivered at Herrischried, West Germany, in August 1969, for friends and supporters of the Institut for Flow Sciences.

present? What brought us to our modern consciousness, to the mentality that is obviously responsible for plunging us into the state of serious crisis we are facing today? A rational therapeutic approach always starts off with a diagnosis; a diagnosis based in turn on the previous history of the illness. Is there, perhaps, an imbalance in our modern consciousness, a one-sidedness which, reflected in practice, proves itself more and more clearly an enemy of life?

Let us, for a moment, think back to the origin of this modern consciousness in the Greek people, in whom it is easy to trace its evolution.

About 3,000 B.C., there lived in the regions later to be known as Greece a folk still capable of an immediate and lively experience of the forces of nature and of the beings underlying them. Let us call to mind Greece's topography: its cliffs, caves, and mountains; the stark contrast between its gentle shorelands and the precipitous formations of its interior with rivers coursing through them, disappearing and as suddenly reappearing; the many springs, tremendous storms accompanied by lightning, and the sound of thunder reverberating through the mountain gorges. One can understand how easily the youthfully impressionable people of that time took leave of their ordinary senses under the impact of such mighty forces and had inner picture-experiences of the creative and destructive powers at work in their environment. Any mental revolving of a question as to whether nymphs or gods of mountain, wood, and stream really existed was simply unthinkable. The still youthful-minded population lived in a picture-world, wholly unacquainted with abstraction, intent upon developing imagination to the fullest and living their daily lives in accordance with it.

Archeologists tell us that around 1900 B.C., there was

an infusion of another element, the Indo-Germans, who came to live in these Greek landscapes among the original inhabitants. They brought with them a far greater leaning toward the abstract. They were quite at home with ideas of good and evil, of purity, of light and darkness. It seems to have come naturally to these invaders to lump together the many landscape deities of similar qualities and gradually apply to them the names we have learned to know them by. Despite a continuing real experience of these divinities and the intermixing of the two groups of inhabitants, which led to the lofty culture of the classic age, there must have been some overall development in the direction of the abstract; gradually merging with the living stream of consciousness of the original population, this accounted for the way the once-lofty gods were brought down to the human level. From this moment on, indeed,these gods were often thought of as possessing qualities "human—all-too-human."

We are all familiar with the further course of these developments. Greek philosophy was born, and the Greek science of the natural philosophers up to the time of Plato and Aristotle. It is perhaps not so generally known that Aristotle developed a many-faceted science that made exact findings in fields such as meteorology and acoustics; the latter were applied in the great Greek theaters.

From this dawning of scientific thought, developments led on to eventuate in the mentality of modern times with its wide-awake but matter-oriented thinking. Along the way great rhythmic alternations took place between the ancient pictorial form of insight and a thin, abstract thought-life; and the battle between them broke out ever and again.

The development of human consciousness to the

point where once-living experience was reduced to an abstract level had to end in disowning that past experience: "For modern man there are only calculable natural laws and natural forces, which can be expressed and grasped in formulas."[1]

How is a mere formula to convey any experience of the wisdom at the heart of things, or how is a coarse-meshed sieve—made to screen out all but crude factual material—to capture the subtler aspects of reality?

The path that has led from wisdom to crude fact is the path that leads from the Tree of Life to the Tree of Knowledge, with all associated consequences for human behavior in the natural and social realms.

Lord Francis Bacon, who lived from 1561 to 1626, demanded that religion and science be separated.[2] From that point on, only what could be measured, weighed, and counted mattered, and the slogan of modern times became "Away with every trace of myth!"

The present inquiry is concerned with the life-element, water. Water, too, has fallen victim to the abstractionist trend, to mechanization. Once the most revered element in every genuine religious ceremony, symbol of the wisdom at work in every phase of living nature, water is now thoroughly "demythologized." It is just "liquid weight," a source of energy, a means of expediting ships and waste matter, a substance suited to running pumps and turbines. Its capacity to drive machinery can be calculated, and this is taken to prove that it is a dead substance. Water becomes just a source of measurable power alongside such other familiar items as pressure, draft, weight, gravitation, inertia, centrifugal force, friction, and nuclear energy.

These energies are harnessed in accordance with the laws of cause and effect and make it possible to construct the machinery and appliances on which our civilization is founded. The goal, the ideal of this development, is the freeing of human energies, in other words, complete automation even in the life-realms, complete automation with its restrictions. These surround human life increasingly, from the cradle to the grave, and have materialized into an environment that we can see and touch.

Before this technology could grow up around us, it had to have become a reality in engineering concepts. Death-related laws had to participate in the thinking of the engineer at his drafting board. A subtle death-process played a godfather role in the conception and designing of the refrigerator, the vacuum cleaner, the automobile, and the turbine. Natural laws were stripped of wisdom and projected into matter.

But this death-process offers a tremendous schooling. Any carelessness in applying the measurements of a formula, the least slip-up in control, can mean violating the laws that obtain in a given instance and can lead to catastrophe. The slightest inaccuracy becomes outwardly evident. Surely the achievement of such absolute conscientiousness is no less valuable as training than are the objects it creates! Absolute honesty, work that leaves no loophole for any sort of pretense—this makes thinking crystal-clear and gives it firm contours. Such thinking takes on matter's stability. Calculation has to be accurate and fit perfectly into the technological spot it was designed to fill.

Even the discovery of the formula inherent in some phenomenon or other requires mental acuteness, a capacity to analyze, to split things apart, to dissect, to separate the essential from the nonessential.

We have been concerned here to show that while technological thinking nurtures exactness, it does so at the cost of engaging in a death-process. The laws operative in a death-related realm are carried along in the flow of living thought. Indeed, a life and death struggle has constantly to be taking place in all such thinking; for maximum economy must be attained, the machine rendered almost 100 percent efficient, and the forces of nature played off against one another.

We still have to achieve knowledge at the cost of including death-processes; the Tree of Knowledge is inexorably linked to death. Should it surprise us, then, that nothing escapes the taint of this death-related way of thinking, from the abstract shapes with which we are surrounded to the policy of straightening rivers? Though leaving nothing to be desired in the technical perfection with which it is carried out, this operation completely does away with the rhythmic alternating of the life element. The last step in demythologizing is the "final clearance" of everything that humanity once experienced as divine.

This thinking done in the realm of death-oriented laws projects an accurate picture of itself in current water developments and can be witnessed in the competitive struggles in modern economic life. It creates attitudes that course through the life-arteries of man and nature; the outer scene reflects the inner.

Nature can stand these developments for awhile, long enough for humanity to complete this schooling of its thinking, but it cannot do so indefinitely. If this type of thinking were to continue to apply its merciless logic too long and infect the social sphere, it would lead to catastrophe wherever life is the thing that matters. And this is bound to occur first of all in the case of water, life's own element.

Technologists and scientists are also aware that catastrophe is in the making, but they do not recognize a specific kind of thinking as the cause. Nor is the right therapy to be found in the "new spirituality" called for at the 1969 conference of Nobel prizewinners and referring mainly to the founding of new universities. That is just where the cause of the fatal developments of recent decades can be sought and found.

A true therapy can only be arrived at by making a correct diagnosis and a corresponding plan of treatment. We should therefore look to see where the trouble lies.

It lies in the postulate that everything in nature can be "explained" by taking recourse to death-related laws. This is simply not true. *There are life-related laws as well.* Thus, modern consciousness contains an inbuilt error in thinking, to wit, that the only laws that obtain are inorganic, causal-mechanical ones. And indeed, a thinking schooled at the hand of what is dead can scarcely be expected to come up with any other kind of postulate after having driven life out of the scene before it.

If you examine any current scientific journal you will find the word "mechanism" in use wherever biological facts are under discussion: the mechanism of cell division, the mechanism of heredity, the mechanism of growth, the mechanism of adjustment, the mechanism of enzyme catalysis, the mechanism of photosynthesis, and the mechanism of the cosmos. Cosmic mechanics! Yet, it was just in connection with the latter that Johannes Kepler, the foremost astronomer of the western world, spoke of a wisdom-permeated organism.[3]

Thus far, our inquiry has been devoted to searching out the causes of the present water crisis, for the therapy we are looking for must be based on a correct diagnosis. We found nature possessing, in addition to its inorganic aspects, realms governed by organic laws that are the polar opposite of the inorganic. It must therefore be possible to train ourselves in a way of thinking in which life courses—a thinking impregnated with the kind of laws that rule living nature. This is not a quantitative problem in the sense of needing more universities; it is a qualitative one.

What changes in practice can we expect to see issuing from life-related thinking? Let us take at least a brief look at the answer to this vital question.

Thinking of this kind would certainly also come to objective expression in the natural environment.

Thus, a truly life-related type of thinking would find reflection in a corresponding landscape treatment. The straightening of rivers and other such life-arteries would cease in favor of returning them to a properly meandering course. Thinking restored to a capacity to take in living formative impulses would be flowing thought and therefore prove hospitable to the flowing, formative forces inherent in systems of life-related laws. Within us, flowing wisdom (such as was made the subject of our previous meeting); outside us, in nature, a water-element that is again allowed to be the carrier of wisdom-permeated laws, a water-element that works as a regulator in landscapes restored to a healthy state of being.

To show specifically what is meant here, let us cite several typical life-factors:

- Life works chiefly as a synthesizer, forming wholes that are greater than the sum of all their parts. Life moves rhythmically, not, however, in a mechanically measured beat. Life is always a little bit eccentric.
- Life is responsive to cosmic rhythms, indeed, entirely governed by them. Therefore, it flows with

the flow of time, in cycles rather than in a straight-line or a mechanical hook up.

- Life always leaves itself free space for maneuvering; it never submits to exact calculation. It moves in cycles wherein metamorphoses and heightenings take place. It is attuned to the cosmos and to endlessness in time and space. Therefore, it is consonant with the highest form of qualitative mathematics, projective or synthetic geometry.

We cannot speak of cause and effect in the life-realm, for there each is simultaneously the other.

Considered from every angle, life is wisdom-saturated. Machines are limited to just one or at most just a very few functions; they are mere "mechanisms." Life can never be fitted into that category.

Now we come to decisive questions: Is there a bridge that leads over from one set of laws to the other? Is there a process that demonstrates the life-function as such? Is there a corresponding bridge between the two realms that human consciousness can use?

There is indeed, and it is water, that symbol of streaming life,—water, every quality of which reveals its life-origin. It is water, where all life-qualities are focused. It is water, focal point of wisdom and hence the medium whereby wisdom is made available to living organisms.

Water's flow constantly links life and death. It is the mediator between the two, and its surface provides a common frontier in nature where they meet. Death is continuously being overcome there.

Human thinking is in a similar position. We spoke of synthetic geometry as suited to the life-realm. It can actually be regarded as an exact method for schooling thinking for that realm.

A moment ago we mentioned a few of life's characteristic attributes. They are the very same ones that could be listed as applying to water, and the same too that characterize thinking when we transform it into living thought.

Now let us look a little more closely at water's characteristics as a border zone.

In the chemical realm, water lies exactly at the neutral point between acid and alkaline, and is therefore able to serve as the mediator of change in either direction. In fact, water is the instrument of chemical change wherever it occurs in life and nature.

In the light-realm, too, water occupies the middle ground between light and darkness. The rainbow, that primal phenomenon of color, makes its shining appearance in and through the agency of water.

In the realm of gravity, water counters heaviness with levity; thus, objects immersed in water take on buoyancy.

In the heat-realm water takes a middle position between radiation and conduction. It is the greatest heat conveyer in the earth's organism, transporting inconceivable amounts of warmth from hot regions to cooler ones by means of the process known as heat-convection.

In the morphological realm, water favors the spherical; we see this in the drop form. Pitting the round against the radial, it calls forth that primal form of life, the spiral.

In every area water assumes the role of mediator. Encompassing both life and death, it constantly wrests the former from the latter.

Water does not constitute a linear divider in the sense of a bar separating one part of nature from another. It alternates between right and left with a living pendulum swing, not only in the sense of creating meandering patterns but in a functional sense as well, thus serving to relate extremes.

Which is to say that water is the primary organ of rhythm, the heart of nature, the element in which we can discern nature's heartbeat. As such it is the polar opposite of a mechanical pumping device: its alternating swing is free. And this eccentricity, this subtle freedom it retains, makes it the element that keeps nature from becoming mechanical, that is, from dying. Indeed, water is the overcomer of the mechanical, and that is why it is so important to imbue thinking with the qualities of water.

We find the watery element present everywhere in nature; it is one of the most ubiquitous of substances. As an entity, it is all-embracing, the repository of cosmic wisdom, which is found in its highest manifestation in the human organism.

Water is like a giant extended over the entire earth, and yet it embraces all wisdom in each tiniest drop. Therefore, it is both a giant and a child, and as such suited to be the mediating agent when the human spirit descends from its diffusion in the cosmos into bodily incarnation. It is indeed true, as science tells us, that human beings come to land out of the water, the element of wisdom. In its first stage, the embryo consists almost entirely of fluids, and only gradually does it solidify the originally fluid organs.

At birth man leaves life's paradisal garden and its streams of living water to find his way into the statics of the solid earth. In its realm, he develops the capacity to walk, speak, and think. He grasps hold of objects and acquaints himself with them. The biography of the individual begins like a journey through a land of hills and valleys, always renewing his or her questioning where the road started and where it leads.

Is there a connection with a higher world, a re-ligion? Is there such a thing as a higher birth into that world?

Perhaps we can understand better now what Christ meant in His conversation with Nicodemus when He spoke of ascending to a higher world through being born again of water and the spirit-breath of air; He spoke of a rebirth that leads human beings back again to heaven, a birth undertaken fully consciously and under the same complex of living laws that governs water.

> Verily, verily I say unto thee, except a man be born of water and of the Spirit, he cannot enter into the kingdom of God. (John 3:5.)

We are of the opinion that the solution of today's huge water problems has never found more precise expression than in the cited talk with Nicodemus. Without a new awareness of water's spiritual nature we will be unable to save this planet that has been given us as our habitation.

The Spirit in Water and in Man

Theodor Schwenk

Conscientious water experts are making every effort to rescue water for humanity. But they continue to work at cross-purposes with other interests. On the one hand, the professionals are warning us that we must wake up to what is going on and develop a new awareness of water as the very element of life. While from the other side, we hear that pollution of the environment with its attendant poisoning of water supplies and the killing of fish is the price we must pay for progress. This has meant that the race being run for the past several years between water pollution and waste-water clean-up has not been decided in the latter's favor. A question hangs over us, unexpressed perhaps but growing constantly more troubling: Is there no authority that we can look to to take the large view of humanity's welfare and to restore order where the wildest chaos now prevails?

No thinking person who entertains this view can escape the clear realization that what is lacking is an organic ordering such as Rudolf Steiner showed to be essential decades ago when he proposed a threefold social organism, in which no one member of the whole outweighed the others, an organism in which cultural activity, civil rights, and economic interests play their roles side by side as freely functioning, independent

---

Based on a lecture delivered at Herrischried, West Germany, in August 1971 for members and supporters of the *Institut für Strömungswissenschaften,* the Institute for Flow Sciences.

spheres. If any one of these spheres gets out of balance, it sets up a self-destructive cancer-like process, and the whole organism sickens. And it is plain to see what member is developing this overwhelming cancer tendency when, in practice, every small and large attempt to clean up water is subordinated to exclusively economic interests.

And what, may we ask, does the above-mentioned "progress" actually consist in? Why, simply in producing goods and services that pander to individual comfort. It is a game taking place in the area of the rational soul, which, at the start, turned with childlike anticipation to enjoying the new technology aborning. In those early days, this required no regulation. However, in modern times, we cannot do without regulating action if we are to avoid the global catastrophe the experts warn about. Humanity as a whole, as well as every single individual comprising it, is being called upon to take the step from a rational-soul to a consciousness-soul orientation. This means becoming aware of the one-sidedness that has developed, and taking healing measures based on an understanding of the fact that the organism of the human race is a single whole. The approach from the whole to the part—or, in other words, grasping the part in its relation to the whole—is a principle that runs, for example, as a life-pulse through the Waldorf School pedagogy to create a foundation in childhood for a lifetime of healing activity.

Earlier steps in the direction of what has now become so vital a necessity are to be found in the existence of such groups as the cathedral building communities, in which everyone who took part, whether as a master or apprentice, stonemason or wagon driver, was caught up in and permeated by the concept of the whole. All the participants were freely motivated to shoulder their tasks, and they worked in brotherly cooperation to bring them to completion. And the temple structure they brought to realization was a physical expression of one already existent in the realm of spirit.

How much stronger and more encompassing consciousness will have to be in order to accomplish a comparable feat within the tremendous context of the organism of the whole human race! How great a contrast with such an orientation is presented by the senseless battling in unbrotherly competition that infects every phase of life today!

In an earlier lecture, we considered the current state of things developed with the evolution of modern consciousness. We described it as a kind of schooling or discipline, and showed how it must now enter a new phase born of a new and different outlook. We traced the development of scientific consciousness as it grew out of Greek philosophy and culture, for these left a legacy that produced strong new shoots at the beginning of the modern age. It was pointed out that human consciousness had to free itself from its erstwhile bonds with Nature's wisdom in order to find its way to its own being, rejecting that maternal ground to advance toward an abstract science for which the very concept "wisdom" had become meaningless. We characterized the thinking error pervading scientific inquiry, which holds that everything in nature, living or dead, can be traced back to lifeless physical and chemical causation.

We then went on to show that quite other laws prevail in realms of life—laws in polar contrast to those obtaining in dead nature; we showed how these can be seen as typically forming wholes and how they function in rhythmic and cyclic processes.

It is always water that mediates between the complexes of laws that govern what is living and what is dead. And so water becomes the great *teacher* at the moment when abstract consciousness crosses the threshold to that other consciousness that once again befriends itself with laws of life.

When we look upon the path human consciousness has traveled as one leading from a state of interwovenness with life's wisdom-working element to an abstract, dead one, a certain question naturally arises. At what point did this transition from the life to the death element in human consciousness occur? It must be possible to pinpoint it, for such marked changes are inevitably accompanied by upheavals.

And it is indeed possible to indicate the moment quite exactly. It falls in the period of transition from the Middle Ages to the dawn of modern times, when religious questioning began to stir western humanity to its very depths.

What sort of questions were these?

They consisted in a spiritual wrestling with such problems as what "transsubstantiation," "redemption," and a "rising from death" actually were. How was it possible to *think* that death can give rise to something living? What is the nature of the risen body? How is matter transsubstantiated? Is it just a symbolic change, a metaphor, a way of putting something, or a material fact? And if so, how? How are bread and wine to be regarded before the change and after it has taken place?

The most outstanding spirits of the Middle Ages wrestled with the question: How can humanity's capacity for knowledge be transformed to serve as an organ for grasping the spiritual realities underlying nature and religion? For thinking had certainly lost its ability to experience such facts; it had lost that earlier way of experiencing in which doubts and questions could not occur because a questioning mentality had not yet developed to shut communicants off from the fact of transsubstantiation taking place in the cultus. Fundamentally, the question was, How can thinking be Christianized? Or, to put it another way, How can it unite itself with Christ? How can communion with Him be made a *thinking* experience?

Let us name here some of the individuals who, as Rudolf Steiner showed, brought the life of the spirit to a new height.

Among those who followed in the footsteps of Plato, Aristotle, Boethius and St. Augustine[1] were Alanus ab Insulis, Herbert of Auxerre, Roland of Cremona, Albertus Magnus, Thomas Aquinas, Averroës, Anselm of Canterbury, Guilbert of Poitiers, Duns Scotus, Petrus Cantor, Simon of Tournai, William of Auvergne, and William of St. Thierry.[2]

All Europe was caught up in the theological questions that were being raised. Realists and nominalists took opposing stands. Councils were convened. There was dramatic controversy, leading to fierce battles, with no quarter given. The problem of "matter and form" was ceaselessly debated. To the extent that works of the Greek philosophers were extant, they were consulted. Arguments grew ever more vehement, bitter and intolerant, demonstrating that the abstract thinking then in the ascendant was simply unable to supply answers to the questions. The *life* element in thought had finally died out. And how, indeed, was dead thinking to achieve any understanding of the life-realm of the Risen One? The decline of thinking proceeded apace to the level of the most abstract sophistry.

And so the great problems were left unsolved. For, as we can see in retrospect, this kind of thinking could not touch them. The most that could be done was to formulate what had once been actual experience into dogmatic articles of faith. No answer could be given to the question of how resurrection, the overcoming of death, was to be conceived.

Now we have reached the bottom of the valley. The descent was by way of science to a conquest of the world of matter. We have chemistry, atomics, and biochemistry, all three of which are approached with the viewpoint that life is just a material process.

Let us listen to one of the world's most prestigious scientists, Professor Paul Dirac of Cambridge University.

> The English Nobel Prizewinner ended his presentation on the fundamental problems of physics with the surprising question, "Does God exist?" The answer was given in close context with the problem of creating artificial life. Dirac did not exclude the possibility of God's existence (despite causality and predestination), but saw it as hinging in a sense on whether the artificial creation of life turned out to be an easy or a difficult job. For if we can do it easily, we don't need God![3]

The questions raised by the Middle Ages in an atmosphere of deepest, heartfelt concern are no longer philosophical in nature; they have become practical questions of technology, and now occupy a level several stories lower than the height on which the Middle Ages entertained them.

Are not the questions confronting humanity and every individual today the very same ones all over again? And have they not become life and death concerns for all humanity?

How can the death processes now taking place in

earth, air, and water be brought to a halt, and matter—water, for example—be restored to life? How can the forces of resurrection, wrested from death, find a locus on earth? And this not just theoretically, but practically, organo-technically speaking? How can we give our cognitive capacity the kind of schooling that will enable it to grasp *the nature of life* and to further it in practical work and action?

These are questions closely related to the earlier questions about transsubstantiation, and they culminate today as they did then in inquiring, How can thinking and how can science be Christianized?

As we seek ways to solve the great life and death problems of human existence, we can connect to that questioning and make use of the fruits of the Scholastic period, even though this cannot as yet provide us with absolute answers. The problem of revitalizing matter depends for its solution on restoring life-forces to our thinking, for thought schooled in the realm of death can only lead to further descent. And to achieve this we will have to develop a new kind of science suited to dealing with the life element.

The problem of rescuing water from death must therefore be solved *inside* ourselves before we can solve it in the external world. When we have transformed the inner scene, the outer one can be restored to order.

And now let us reiterate, Can we progress from the present into the future where knowledge and nature are concerned?

Yes, we can! In nature, plants demonstrate how to do this as they take up and enliven dead matter in the soil. This is something human beings cannot do in their bodily organisms; that is why they must ingest living matter from the plant world.

How do plants do this? They have to depend on water to help them. So we must ask how water enlivens matter in conjunction with plants.

Water is able to accomplish this because, as we have shown before, water is itself the fluid life-process in its every aspect. Thus, it constitutes a kind of primal bodily organ of the life element. It occupies a middle position between life and death. Everywhere we look in Nature we see water setting up crossing points by inserting itself between opposites, for example:

- Between the acid and the alkaline in chemical processes, thus mediating material exchange;
- In the realm of gravity and levity;
- In the realm of warmth, between radiation and conduction, as convection;
- In the realm of light, between brightness and darkness, as seen for example in that primal phenomenon, the rainbow.

Working thus at the balancing point in natural processes, swinging back and forth like the lever of a pair of scales, water serves to maintain equilibrium. It is the omnipresent, ever-reestablished active element in life's freedom-zone.

But now we need to ask a further question: How does water manage to insert itself so wisely *between* opposites?

It can do this because it seeks nothing for itself on either side of that boundary where two extremes impinge on one another. Formations that come into being along such borders—involutions, for example—are swept away again at once so that the mediating of new patterns can take place. Water thus brings about those two archetypal life-processes—becoming and subsiding. It must be apparent that wisdom is built into such attributes. Wisdom inheres in things by virtue of the fact that water is present in everything alive.

Now how are we to picture the transition from deadness to aliveness in human cognition, since human beings are not organized to achieve this in a *bodily* sense? How can we bring what goes on physically in plants into the realm of our thinking activity?

Let us take an example. Let us imagine that we are standing in front of a blackboard that has drawings on it. But the time comes when those drawings that were there at first have to be erased in order to make room for new ones. That is how our soul-life functions. The same thing happens when we go to concerts. We erase the great variety of impressions left in us by everyday living and make our psyches over into comparatively unwritten slates in order to take in new impressions and to do justice to them.

What significance does this kind of practice have for our capacity to listen in conversations where something is being communicated? To the degree that we achieve real listening—freeing ourselves from all other thoughts—what lives in the soul and spirit of the other person is revealed to us. Physicians, indeed, have to carry this process one step further, for they must try to live into even the physical state of the patient confronting them, relying by no means just on what they hear the patient saying. We might simply call this kind of relating to a fellow human being renunciation, a renouncing, too, of overhasty judgments and preconceptions, along with the practice of quite a bit of patience.

These are some of the qualities of water that enable it to serve wherever it occurs as a quickener and healer. To the extent that we free ourselves from self-absorption we too can be transformed and quickened.

What we are contemplating here is something in the nature of a true cultus, a cultus wherein renunciation is the sacrifice or offering—openness and willingness to receive the imprint of a fellow human being, transsubstantiation, illuminating insight into the other's being, and spiritual communion with him or her.

To entertain thoughts like these in a time beset with the burning problems involved in rescuing the elements may, at first glance, seem anything but practical and realistic. But it is nevertheless basic to their solution. There can be no such thing as a quickening of matter, no communion, without prior sacrifice, which in this context means renunciation.

Let me put it slightly differently: To serve the cause of water adequately and to do something practical for the element as such, we must get to know it by taking in its true being. And how do we do this? Why, by treating it in the very way exemplified by its own behavior, that is, wherever we encounter it, we wash the tablet of our souls clean of all other impressions in order to allow the being of water to make its imprint on us. If methods of this kind were gradually to reestablish themselves in scientific practice, we would not have long to wait for the development foreseen by Rudolf Steiner when he spoke of the laboratory table having to become an altar.

Basic to the path of knowledge taught by Steiner is the extinguishing of preconceptions, prejudgments, and pre-images, in the sense that to develop imagination, everyday impressions are erased; to develop inspiration, the previously held imagination is extinguished; and to develop intuition, that is, the immediate perception of another's being, the prior inspiration, is obliterated.

This does not call for the bodily asceticism practiced

in the Middle Ages; it simply requires application or the energy to develop insight into the inwardness of things. It does not seek to control nature and to wrest her secrets away from her; she gives them to us gladly if we are interested in her rather than ourselves. When this happens, the technician and the scientist are on the way to a technology of life.

Now how do we go about re-enlivening water? There is no other way than to develop what the professionals themselves describe as "a new water consciousness."

So long as the other two members of the social organism are subordinated to an over-proliferation of economic life, the lofty, selfless being of the watery element will remain a stranger to us all. We must take to heart words recently spoken by two insightful hydrologists:

> People have not yet entered into a right relationship with technology. We must learn to think in a different way and change that relationship. Man, whom all our efforts—including those in the field of technology—should serve, must be made the central focus of this new way of thinking. . . . Technology's ingenious accomplishments are the product of the conscious human spirit. But human intellect is not by any means our highest function. Despite the wealth of its ingenuity and its logic, it lacks depth of feeling. It helps us determine degrees of efficiency and guides us to economic applications, but it has nothing to offer on the score of "good" and "evil"; it does not motivate us to resist the bad. Intellect permits the misuse of technology in carrying on wars, in greed for possessions, and in other evils. Trends like these can only be fought by an inner soul-change on the part of every individual.[4]

The late Albert Betz, an engineer who accomplished such great things in aerodynamics, once said,

> Seeing what people have done with what we engineers have created for them certainly dims one's pleasure in the accomplishment. One has to ask oneself whether we were right to pursue such research and to do such creating. The answer has, of course, to be in the affirmative. But one thing has been overlooked: we have got to take the greatest care not to let what we create get into the hands of people who lack an ethical maturity commensurate with technological advance.[5]

The questions we have raised as to the possibility of re-enlivening the earth's water supplies brings us to the threshold of a technology based on the complexes of organic laws, a technology attuned to life itself. Does not such a technology particularly require us to "take the greatest care not to let what we create get into the hands of people who lack an ethical maturity commensurate with technological advance"?

Indeed, the questions arising out of the current situation are in essence ethical, not technological.

Today's environmental problems are clearly recognizable as newly resurrected spiritual questions that have become matters of life and death for present-day humanity. They cry out loudly, demanding solution after so many centuries, solution with new human capacities.

The consciousness of humanity as a whole has completed its descent into earth and the kingdom of dead laws. Now it has become the obligation of the individual—the "needle's eye" of the human race—to travel the road to the realm of life, to a rebirth learned from water's being.

Water: Destiny of the Human Race

Theodor Schwenk

The Crisis

A look at history teaches us that water has not just recently begun to affect humanity's destiny, but has always done so. All the great civilizations along the Nile, the Euphrates and Tigris, the Indus, the Huangho owe their origin and their flowering to the watery element. It seems self-evident that water is the most vital factor in the existence of a culture.

As we discuss the theme set for today, we will need to cast a little more searching light on inner aspects of this seemingly obvious fact. For though humanity has reached a peak of civilized ease in our day, we see beyond the shadow of a doubt that this achievement has been accompanied by a swift cultural decline that leaves what we used to take for granted by no means certain.

The water crisis of today has grown to planet-wide proportions, both in its quantitative and its qualitative aspects, so that the whole human race is now actively as well as passively involved in it. Looked at from the therapeutic standpoint, the facts demand that where water is at stake humanity is going to have to conceive of itself quite consciously as a single whole, and begin equally consciously to take fitting care of its dwelling

---

Based on a lecture delivered at Herrischried, West Germany, in August 1973, to the supporting members and friends of the *Verein für Bewegungsforschung* (Association for Motion Research) and the *Institut für Strömungswissenschaften* (Institute for Flow Sciences).

place, the earth-organism. The only question is how to go about it.

The situation that we are now about to illustrate is not due to some vague causation or other; every single human being has had a share in it. It should therefore be possible, by involving everybody, to find a solution to the problems that currently beset water. And though this may seem at first hearing an impossible demand, it is nevertheless one that appears directly or by implication in reports throughout the civilized world of water catastrophes. The following examples will serve to bring home to us particularly how widespread the problem is and how the individual is being appealed to:

> According to the latest research findings of the professor of pathology at the University of Kumamoto, hundreds of Japanese will be undergoing the experience suffered by the mother of little Yaeko. One of these days they will fall ill with a fatal environmental illness, the germ of which they are carrying. In the majority of cases, they will die after a long decline, for there is as yet no known remedy. The reason for the professor's research was the very recent discovery, on the coast of Ariake in Japan's southern islands, in a region not far from Nagasaki noted for its natural beauty, of a new epidemic outbreak of minamata, an illness caused by industrial waste-water containing methyl-mercury. It is contracted by eating fish caught in mercury-contaminated waters. It paralyzes the nervous system and leads to deafness, pain, crooked limbs, distorted vision and hallucinations. . . . Professor Takeuchi . . . found that ingesting even miniscule amounts of mercury over a period of time can lead to illness. Young children and the elderly are especially vulnerable. The professor therefore fears that in its earliest stages this malignant disease could already have spread over the entire country.[1]

> Japan's industrial growth is no longer a matter regarded with awe and reverence. One of its largest chemical concerns is currently threatened with bankruptcy as a result of the courts' deciding that it must pay for all the environmental damage caused by its operations. The firm of Chisso and Company, manufacturers of plastics, has been adjudged guilty of polluting Minamata Bay with its methyl-mercury contaminated waste-water and thus responsible for the illness called "the Minamata disease" after the body of water it comes from. . . . When, last July, six chemical companies were required to pay damages totaling 880,000 German marks for having been perpetrators of the scarcely less horrifying "Yokkaichi asthma," the president of an influential trade organization expressed industry's attitude toward this judgment as follows: "Sentimental sympathy with the minority who have suffered can only lead to disastrous consequences for the growth of Japanese industry."[2]

We could cite any number of similar examples from almost every country on the earth. Pollution and the poisoning of soil, air, and water are on the march and laying hold on every land like death itself, which spares nobody. Even remote Lake Baikal, the largest fresh water reservoir on our planet, is no exception:

> The huge lake, lying 636 kilometers long between mountain chains that soar to heights of 2000 and 3000 meters, shimmers against a background of ancient firs and larches. On the opposite southern shore some 40 kilometers away, a heavy brown smoke-cloud hangs over the blue waters where the largest cellulose factory in the Soviet Union has conjured up the specter of environmental ruin over crystal-clear Lake Baikal.[3]

It matters little whether we are considering Lake Baikal, the Mediterranean coast, the open Atlantic, the world's rivers or the earth's ground waters. These great life-regenerators are all similarly threatened or already struggling in the throes of death.

The piling up of facts like these has to mean making constantly repeated appeals to human conscience. But as appears from the examples cited, ethical reserves have been exhausted. New ways must be found, ways that awaken reverence for the life-element.

If we are to find the way to deal therapeutically with the current situation, we must first look at what has brought it about. Medical doctors inquire into the previous history of a patient's illness before they feel they can proceed to treat it. What, we must ask, has caused these problems of water quality and quantity to develop?

Discussion centers around two major causes: overpopulation and industrialization.

The Disease of Uncontrolled Growth

Let us examine the latter more closely. It began with the individual. A craftsman who made some article for which demand was growing enlarged his premises. Sons came along and expanded what had begun as a small business. Thus, family firms with famous names developed; the story of many such concerns is thoroughly familiar. They go on growing from generation to generation. There are amalgamations, industry-wide combines, conglomerates, and so on. Since these are not isolated instances, but rather a process that spreads out over the map in all directions, networks come into being and, with increasing competition, become areas of tension—all over the earth domains, small to begin with but growing ever larger, each one eyeing the others in the race. So we see impressive growth, in some cases leading to the creation of gigantic cartels, but lacking any overarching regulative principle such as inheres in every organism worthy of the name. The process goes on in uncontrolled burgeoning, to the point where markets "have" to be artificially created for the sole purpose of earning further profits.

If something analogous were to happen in an individual human organism, we would describe it as a proliferation of matter at the expense of a catastrophic loss of formative capacity, meaning thereby a disease process of the times that afflicts rich and poor alike—cancer. A reflection on a small scale of something much larger that happened long ago! We might call it a lack of inner coordination in the functional parts of the organism. Way back at the beginning of the century, Rudolf Steiner called attention to this process as one portending catastrophic consequences.

The symptoms of this process have in the meantime become noticeable enough to be frequently mentioned and described, as the following quotations demonstrate:

> Is our prosperity killing itself with overdoing? Help will have to be sought in the long run in restricting industry to needed production, not in a course of unbounded growth. This insight is vital to success in achieving a balanced social and economic system in which consumption is bridled and energy requirements cease feverishly climbing.[4]

Concern for the future:

> Society seems unable to recognize the fact that lack of moderation must end in catastrophe; production and consumption cannot go on increasing while the environ-

> ment's need for protection—especially in the matter of keeping air and water unpolluted—is overlooked in our egotism. Gabor made the statement that industrial growth will therefore have to be throttled down.[5]

Questions of environmental protection:

> The question as to how our economic system can be harmonized with what must be done (for the environment) was the theme to which the University of St. Gallen's symposium on economic and legal aspects of environmental protection devoted itself. . . . Professor H. C. Binswanger saw the real problem as lying in the economy's compulsion to expand. Thus the desire for steady growth leads to an annual increase in the gross social product, but bypasses the question as to whether the loss of natural resources justifies the increase. In view of the consequences in areas such as energy consumption and the setting of tolerance limits on permissible pollution of the environment . . . the question arose whether the economy could continue functioning under these restrictions and what alternatives there might be. No practical proposals were advanced, however.[6]

This summing up of the St. Gallen symposium puts its finger on what the main problem really is and raises the question as to why no practical proposals could be made.

After all the happenings that led up to these quoted reports it should not be difficult to supply a summarizing answer: no overarching design has been devised to meet the situation as a whole, and is altogether missing. That is why we are always hearing a demand that an international authority be set up and empowered to act in the total human interest with the issuance of directives creating order in the situation.

Instead of this demand being met, however, we all see daily how entangled economic considerations, civil rights and cultural interests are. In the East, politics dictates to the economy and the cultural life; in the West, the economy dictates to the cultural and rights spheres. The result in both cases is a caricature of the "composite king" in the Goethean fairy tale, who collapses at the end into a pathetic heap of metallic dust.

Healing Design or Cancerous Proliferation?

A design for a healthy social order has been available in the form of Rudolf Steiner's idea for the threefolding of the social organism:

> Equality for all in the area of rights,
> Fraternity in the economic sphere,
> Freedom in the cultural life.

Before the old outlived social forms were swept away by rivers of blood during the French Revolution, there was a brief moment in the history of the world in which an ideal for the future rang out:

Liberty, Equality, Fraternity!

It made its reappearance some time ago, but in its new aspect it had taken on form and related each one of its three challenges to its proper sphere, so that a true social organism might arise out of the chaos and confusion of mixed elements.

We are engaged in searching out the causes of the present water situation as it exists all over the earth, and can at least ascertain definitely that they lie in problems of societal relations in our modern industrial society.

Who would have thought, or would think now, on

hearing Rudolf Steiner's assurance that the threefolding of the social organism would come in the form suggested—though perhaps ushered in by frightful catastrophes—that it might be natural catastrophes of some sort that would present humanity with this alternative as a saving life-element! Indeed, who would imagine that a natural catastrophe could be precipitated by wrong developments in the social structure—for example, by a cancerous proliferation of the economic life in which the "struggle for existence" rather than brotherhood has been raised to the status of a scientific and practical idol?

We can and must, of course, go a step further and inquire what has made people with social concern incapable of taking in the creative idea of the threefold social organism and bringing it to realization. The answer—again, given by Rudolf Steiner—is immediately clear: it stems from the way schools teach natural science. The concepts of man as the highest animal and of the "struggle for existence," made so much of in museums and disseminated in publications by the millions, cannot fail to have real consequences since every image we form has the tendency to become reality. Every crime, even the craziest fantasies witnessed in the movies or on television, presses to be realized. These are facts confirmed in courtroom hearings.

Having been educated to it, we all know and daily experience what the elements of scientific method are: analyze, separate, split up, and take apart, until we end up with atoms or cell nuclei.

The great scientists, men like Hermann von Helmholtz, Michael Faraday, Gustav Hertz, Charles Darwin, Sigmund Freud, were still influenced by tradition in their own need to see things whole.[7] They could feel enthusiastic about art and religion and cultural values. But their successors find as their inheritance mere parts and pieces of the great artist Nature: atoms, chromosomes, and nucleic acids.

Thus, wonderworlds reveal themselves in the course of inquiring into nature's single facets, in the process of snatching her secrets away from her, of approaching her in attitudes bred by the struggle for existence, and trying to get the better of her. How does the application of this kind of knowledge to technology, the servant of the economy, look? Is it applying the scientific method to itself in the sense of practicing self-knowledge, as we are attempting to do very briefly here? What is it doing to the totality of things when, in its attention to detail, it fails to see the whole organism of the earth and of humanity?

The facts cited at the beginning of this talk are clear evidence that science has overshot its goal and lost sight of man, the wholesome measure of all things. Warning voices are being heard from its own circles.

The separating that has been going on should be succeeded by putting things together again, first in the scientific approach, and then, as a consequence, in technology and economic life. Scientists and technologists have obviously forgotten the cardinal question: How does nature create wholeness? Ideas are everywhere to be found in her; if it were not so we could not rob her of them as we do. What being are we confronting in this world of ideas and order? Why do scientists call a halt to their questioning at just this point? Have they been so affected by their analytical schooling that they simply cannot get themselves out of the rut again?

Therapy

This is where the prospect opens on healing possibilities. But let us recapitulate briefly the sequence of steps we have taken thus far.

An analytical approach to knowledge stamps its imprint on thinking, lives itself out in technology and—as the struggle for existence—in economic life, with resultant cancerous tendencies and a battling for power. This kind of thinking destroys nature's wholeness.

This in turn covers the whole planet with a network of conflicting egoistic impulses responsible for losing the total picture, losing the possibility of regulation, losing the bases of life—indeed, losing the life-element itself.

To recognize these facts, however, makes it possible to indicate the direction in which an effective therapy should go:

- Foster wholeness in the life of knowledge;
- Foster organic thinking, schooled by the life that builds whole organisms;
- Foster a thinking capable not only of separating but also of relating things, a thinking with a round, inclusive view that takes in the whole circle, rather than the linear shooting at targets that is generally practiced at present. Such a view would lead to self-discipline through awareness of the whole.
- Foster brotherhood and reverence for man in economic life. All this must become the business of humanity, counteracting the Darwinian doctrine.
- Work at structuring the social organism in accordance with the architectural principles of the threefold order.

Wholeness

Now we must proceed to illustrate these aspects with a few examples, taking water, the element to which we devote ourselves here, as our guide in dealing with questions of wholeness. This is a role to which it has always been particularly well-suited.

Water accounts for most of the wholenesses, forms and shapes in nature's realm, and most especially in all living organisms in the plant, animal, and human kingdoms. Indeed, water makes the earth organism one single whole, functionally at least. The earth organism has distinct functional organs, which, though they are not yet recognized as organs, show themselves to be such as soon as they are studied in a larger context. Since all of earth's life depends for its existence on an interplay between earth and the cosmos, we are justified in looking for such an interplay where its organs are concerned. We need not raise the question here as to whether gravitational forces are the only ones that play a decisive role in the relationship of earth and cosmos; it is enough to know that the cosmos is symptomatically "at home" in the water-element.

Coastal dwellers experience the rising and falling of the tide as a pulsebeat. Such tidal "waves" making their way north from the southern tip of Africa, for example, roll up through the Atlantic Ocean, pass the equator, and show up in European waters along the coast of Cornwall, in southern England. Making their way further north between the British Isles on the one side,they circle around the northern part of England and then move south into the North Sea, where they come together again with the other tidal current that has branched off to make its way counterclockwise around England through the English Channel. If we follow the course of this tidal front in an atlas, we have before us the phenomenon of a "pulsating wave" which, as we know, has been found to depend on the position and movement of the moon in relation to the earth and sun. Indeed, the concepts "spring-tide" and "neap-tide" embrace just this cosmic correlation. These tides are like a constant pulsebeat that makes itself felt throughout the earth's entire water sphere and causes it to be a single entity.

Relations to the Laws of Music

These "pulsations" can be followed right down into quite small dimensions, cropping up again in lakes, for example. Just as every bowl filled with liquid has its own vibration that varies with its size, shape and depth—waiters have to have an exact feeling for this—so does every body of water, say a lake or a bay, have its own characteristic pulsation. The Swiss physician and investigator Auguste Forel made the most exhaustive study of these pulsations in Lake Geneva and described them.[8] They are called forth and influenced by changes in air pressure and tidal elements. The Lake of Constance, for example, has a "pulsation wave" that goes down the lake and back again in about an hour. If we consider the fact that every lake and basin has its own special vibratory form, conditioned by its size, shape, and depth, we can speak on the one hand of a kind of individual "tonal color" and on the other hand of a definite tonal pitch in relation to a basic prime. Let us picture what happens when the moon, traveling over the earth from east to west, passes over earth's various lakes. It sets the complicatedly formed Chinese lakes pulsating, for example, or we could say, it intensifies their vibration. Then it passes on to Lake Baikal, and then to the Aral Sea, the Caspian Sea and the Persian Gulf, the Black Sea, the Red Sea, the Swiss lakes, and after these to the lakes and bays of Ireland. But now we must add to this the fact that pulsations or vibrations of this kind follow exactly the same laws that govern vibration in wind instruments, flutes for example. These are rhythmic movements lying below our hearing threshold, but nevertheless present and actively setting bodies of water vibrating. Thus, we may speak metaphorically and say that the moon, traveling on its orbit around the earth, "plays" on all the various bodies of water, from east to west, and they all "resound," each in its own individual resonance and tonal style. Since fairly large bodies of water are involved here, these vibrations last correspondingly long, and are subject to ever renewed stimulation by the moon. Can we not talk of this in musical terms and say that a multiplicity of single instruments has been joined in an orchestra that plays its score loudly or softly day by day, year in, year out? It would be a fascinating task to transpose into audible tone the melody thus produced (though it might sound unaccustomed indeed to our human ears).

Yet these lawful patterns have been built into the microscopic space of our auditory passages. Let me give another brief illustration of the process.

It sometimes happens that the Cornish coast is visited by very unusual, long waves in addition to those of the daily flood tides, and they break against the cliffs there. These waves often come from the Cape Horn area, at the southern tip of South America. We can calculate the time of their origin in some mighty storm in that vicinity. What has happened is this: The sea in the storm area is churned up to such an inconceivable degree that any kind of order is ruled out. All sorts of different wave-systems come into collision and tower up, creating deep troughs. However, the innumerable smaller waves gradually give way to larger ones, which have the peculiarity of rolling much faster. The latter outdistance the lesser ones, rushing toward the coast of Cornwall as majestic billows. One would wait in vain there for the arrival of the shorter waves, for those have long since been left behind; they have flattened out and disappeared at a distance from their place of origin that varies according to their size.

Brilliant researchers have discovered this very phenomenon, which takes place over such extended areas, in processes occurring in the fluid of the inner ear, in the cochlea. This process starts at the fenestra ovalis, a membrane on which rests the middle ear bone known as the stirrup or stapes. This area, which encompasses the middle ear and the ear drum, is where vibration is set going in the ear.

So that when the stirrup vibrates and activates the fenestral membrane, it starts a wave there that travels up the cochlea—a wave which, if it is long enough, presses right up into the dome of the cochlea, whereas the small waves, corresponding to the higher tones, have already stopped moving near their place of origin in the neighborhood of the fenestra. Georg von Békésy, the researcher to whom we are indebted for these findings, discovered that the low tones—in other words, the long waves—are localized at the far end of the cochlea, while the high tones with their shorter waves remain close to their starting place.[9] We could, of course, conceive here of processes like those in the Atlantic Ocean being projected into the inner ear; but we can just as well reverse the picture and see in these processes a gigantic sense organ of the earth that enables the most remote regions to communicate with each other by means of ocean waters and carries "information" to them. We see it extended over the entire planet, like a consciousness that links and makes a single whole of the closest and remotest parts of the earth.

Everyone can undertake for himself a further study of such matters as, for instance, the role played in the circulation of warmth and substances by the submarine ridge in the Atlantic between Iceland and the Faroe Islands, or the extraordinarily interesting metabolic process in the oceans that have been described as a veritable domestic economy of material contents.

One thing becomes clear from such a study, that is, that the qualities of water are such as to make it the basis of all organisms, regardless of size, and that water therefore meets all the requirements for forming organs, which it then links together in entire systems. Our handful of examples will have sufficed to demonstrate this threefold functional organ forming of the earth: it possesses the system of organs necessary to digestion, to rhythmical "life" and to the functioning of the senses. We confront here a great macrocosmic human being, interrelated with cosmic processes and weaving a wisdom-permeated fabric of internal and external functions. In earlier times people must have had direct experience of this being, which they called "Adam Cadmon."

Water's Role in the Symphony of Weather

In speaking of the earth as an organism, that is, as an organic entity, it stands to reason that the basis for this must be found in the water-element, the basis of all life. This is amply demonstrated by countless examples that mutually support and uphold one another. But the earth's atmosphere too shows itself to be part of this organism, this great functional entity characterized by orderly arrangements of spatial and temporal elements. So we speak of the planet's atmospheric mantle, of wind and weather processes, of the seasons and their changing warmth and cold conditions, of light and darkness, of sunsets and sunrises glorious with color. But we experience these ever-recurring, ever-varied processes as manifestations of one single whole, and must be quite clear that they could never take place were it not for the presence of water in the atmosphere. The mighty rhythms that are manifestations of the planet's life are mediated, as all life is, by the life-element water. The technical way of putting this is to call water the "motor" of all meteorological processes. But it can also be expressed thus: Water is as vital to the regulation of climatological and weather processes as oxygen is to the breathing processes of living creatures; without it, everything would become a desert.

The familiar process underlying the buildup and disintegration of the great stationary high-pressure and low-pressure areas—the Asiatic High and the Icelandic Low, for example—is such as to make one conceive of it as a mighty breathing of the continents that sucks in and then expels the air above them in connection with the annual course of the sun. But interspersed in this regular breathing, like swiftly changing, marked variations on the theme and *between* the stationary mountains and valleys of air pressure, we find the winds of the wandering low-pressure areas or the highs with their dramatic manifestations of warmth and cold.

There are air masses containing air from the subtropics at one period, at another air drawn from polar regions, and on still other occasions damp oceanic or dry continental air. Even in the early days, meteorologists felt called upon to characterize cyclones, the low-pressure areas most familiar to us, in a special way that we ourselves could not improve on. We have already quoted Paul Raethjen's statement that cyclones have a typical life history and that atmospheric phenomena must be viewed as aspects of a single living whole.[10]

It is particularly noteworthy that the leading meteorologists, ever freshly reminded of the totality of the earth-organism and led again and again to its contemplation, should find it necessary to resort to terms borrowed from biology and music theory to illustrate and point up processes occurring in their field. And we must count among the most significant findings of meteorological research the works of August Schmauss on the lawful patterning of the "dynamic year," which he discovered and to which we will need to refer here. Schmauss takes as his standard of measurement for the dynamics involved in European weather the difference in air pressure between St. Mathieu at the northwestern tip of Brittany and Lerwick in the Shetland Islands, the almost northernmost point in the British Isles. This connecting line may be described as the door through which weather enters on its way to Europe. High and low points occur in these dynamic developments. We are familiar with some of them, for example, the May drop in temperature. Schmauss writes:

> The "dynamic year," as we might call it, usually begins on the 29th of September with a bottom low along the pressure gradient from St. Mathieu to Lerwick and with a minimal atmospheric activity familiar to many as a phenomenon of Indian Summer. From this point on, the gradient—and the zonal circulation with it—increases in a series of waves until it reaches the winter high, falling on the average on January 9. The waves . . . are the

> expression of a battle between summer and winter, from which deep winter emerges victorious. But the culmination of its activity is also the turning point in the battle. Here winter begins its retreat with a series of rear-guard actions that show up in wave-forms in the meteorological curves. Activity decreases until it reaches the annual low-point, which occurs on the average sometime between May 22 and the 6th of June. This stretch of days quite often brings the first heat wave, for there is not much movement in the air, and as the sun approaches its zenith it can therefore make its radiation fully felt.
>
> At the beginning of June this period of especially pleasant weather often ends quite suddenly. So-called summer monsoon weather starts . . . with heightening atmospheric activity. Cooler west winds pour in, frequently bringing with them along the wide front thunderstorms like those of India, the classic land of monsoons. The gradient remains high, though with wave-like weakenings, until September 16. Then there is a sudden plunge to the autumnal minimum, matching in pace the swift upward climb from the spring minimum. The calendar of events is thus by no means a matter of chance; instead, it actually deserves to be called a *collective timetable.*[11]

Findings such as these were the outcome for Schmauss of decades spent observing and measuring. And they bring home the fact that at the equinoctial periods as well as at the sun's turning points it is often possible to observe the interesting phenomenon of a mirroring of curves in the pressure of the air. In other words, the air pressure pattern prior to these times finds itself reflected in the pattern that follows. Time's course thus brings to expression exactly the same orderly patterns that are found in music. And when Schmauss points to Johann Sebastian Bach's musical compositions in this connection, he is merely recognizing higher laws in both these and in weather processes. In certain places in his compositions where Bach wishes to indicate a working in of higher forces, he makes use of just such musical mirrorings. We know too from the research of Rudolf Steiner that the course of spiritual events often takes place in a pattern of which earth-happenings are a mirror.

Schmauss goes on to say:

> A meteorologist familiar with these facts sees in this scheduled taking-of-turns by masses of air something in the nature of an orchestral score that indicates where an instrument comes in—though on occasion it may miss its cue. Even so, it is a great pleasure to be familiar with the score. . . . For the meteorologist . . . who is familiar with the tremendous changes in weather patterns, it is indeed an experience to behold the structure that underlies them.[12]

In his work, *Biological Thoughts in Meteorology*, Schmauss gives a detailed characterization of the structure of a low-pressure area:

> If we want to understand the structure of a low-pressure area, we have to take a great many examples of depressions and then try to form a picture that embraces them all. Our knowledge is quite equal to defining the type "low," but that is a collective term. Single lows are *individuals,* evidencing variations often brought about by very slight changes in the prevailing circumstances. We might, borrowing a musical term, call them variations on a theme. We marvel at the ingenuity of a Bach or a Reger who could write any number of such at will. Our atmosphere is certainly a master at this too.[13]

In relation to the meteorological phenomena that take place in our atmosphere, we therefore have to speak of time-structures, of an organism, a totality, a symphony or score from which the processes involved receive their patterning. And in the case of cyclones we have to speak of individuals.

Does not nature herself break into speech here? And does she not move us to inquire: "Who is playing the score, then? Who is the composer, if the compositions we encounter here abide by the same laws that govern music? Who composed them with wisdom enough to enable life on earth to flourish?"

It must strike us that the main accents of the dynamic year coincide with the chief festivals of the Christian calendar—or, vice versa, that, for example, the dynamic high in outer nature takes place at the time-point that marked the original Christmas celebration; that the low, coming at September's end,ushers in a fresh new impulse that carries the dynamic to its high in the early part of January. Or again, that Whitsun, a movable feast dependent on when Easter falls, is likewise reflected in what happens in the weather. Is not all this to be attributed to the being who is to "reappear in the clouds," the being who in past days of creation composed the life-functions of this organism in accordance with spiritual laws? The cosmic man extended over all the functions of the outer world is to be found in meteorological processes too. Or we could put it another way and say that what is spread out in the surrounding world is found again in man-the-microcosm, contracted into his physical body. (Need we be surprised to find the great and small going hand in hand and to see so many hitherto little understood interrelationships with a bearing on our meteoro-biological susceptibilities turning up?)

The Nature of Water

Neither the cosmic man in the world outside us nor we small human beings could exist without water. We are all actually condensations, as it were, of the watery element—a state of affairs that could not have developed if the whole human being were not already functionally preestablished in it, meaning here the functions of the metabolism, the circulation, breathing, and the sense organs in the waters of the earth.

So we may say that functional man can be discovered in the great world around him in three different manifestations:

1. in the water element itself,
2. extended over the earth-organism,
3. in the atmosphere.

What is the nature of this element in which the whole human being already exists in prototype?

It is an element capable of serving as a medium for the action of every kind of force, even the smallest, and of their polarities:

- It occupies a position between life and death;
- Between gravity and levity: we owe the possibility of thinking to the brain-water's buoyancy;
- Between light and darkness: their interaction gives rise to the rainbow, the primal phenomenon of color;
- Between base and acid, in a neutral realm, where it mediates all chemical change;
- Between stillness and movement, whose interplay forms the basis for every rhythm and every pulsebeat;

- Between solid and volatile: in other words, fluid, that is, endowed with an unlimited capacity for form-creation and metamorphosis.

In the case of beings endowed with moral consciousness one would have to say that such a fullness of possibilities can be the fruit only of a complete renouncing of everything fixed and settled. But this is just what makes water the transforming element it is and gives it the ability to receive the impress of higher elements and enter into combination with them. The three stages: sacrifice, transformation, and communion, are inscribed into its being, and they underlie its every action, as we can see wherever it confronts us.

Through a knowledge of this element out of which we were once physically born, we can attain, in a higher schooling, to a further, second birth on a higher level.

Rebirth out of the *spiritual* being of water: the circle closes. Leaving behind us the living waters of the Tree of Life, we had for a time to experience death beside the Tree of Knowledge, to the point where our actions brought about the present situation in which the life element, water, is involved. That situation must now become the means to our awakening and asking questions about the Tree of Life and the new Adam.

What is it in the watery element that turns its gaze on us?

Why our own being as it must become! But this is to say, our destiny.

What is it in the whole earth's water-organism that is looking on the whole human race?

Why, the being that we may call the spirit of both earth and humanity—man's archetype, Adam Cadmon!

And what directs its gaze on all of us today in water?

Why, what we have made of it, our own being as it has developed.

We have become water's destiny, and from this point onward water becomes ours.

At life's primal spring the Norns spoke thus:[14]

> Your own is your all,
> Your harm as your healing,
> Your will and imagining,
> Musing and being. . . .

This holds true for the whole human race. What humanity has made of water's organism, of its own higher being, in brotherhood or what goes counter to it, looks out at us.

Man as Co-Composer in Destiny and Design

We have tried here to make a diagnosis and to indicate treatment. Now let us make a prognosis also—if the patient will cooperate by including water's spiritual being in his understanding and acting out of a will guided by it.

Everybody can move in that direction by creating order in his spirit's household and by tuning his instrument to the entirety of mankind, sacrificing narrower intent to a wider view that looks back and into what lies about him or her.

That can be achieved by the healing being that represents the social body of humanity in its wholeness. Healing means "wholing"—reestablishing the design, the image.

The Norns put it well:

Your own will is your all,
Your harm as your healing,
Your will and imagining
Musing and being.

We form with our fingers,
From storehouse eternal,
The twirled threads of life:
Each man's single lot.

We're spinning and spooling,
We're winding and weaving
Tapestried deeds
On the loom of the world-all.

Enwoven for ages,
From us is the warping,
Your own is the woof,
The designing, O Man![15]

The design they speak of is one's personal life. But what is being woven today is the pattern of the destiny of the whole human race.

The new commandment given by Him is brotherliness, in which the human race becomes one whole.

Through each one attuning himself to that whole, humanity's destiny becomes another "net" woven over the entire earth, its pattern an interweaving of brotherly elements.

The present water situation delivers the warning that every one of us is implicated, and every one responsible for the outcome. That alone is the solution of the problem.

So we may ask: "Is everyone then the final authority so often sought for? Is every single individual a co-composer of the great human score, the threefold composition in accordance with which our planet can be properly cared for?"

Everyone has the opportunity, everybody the responsibility.

Everyone has the possibility of playing the instrument of his or her own being and can share in every voice of the mighty score by tuning in to the orchestral whole.

An entrance may be missed once. But what a pleasure it is to become acquainted with the score, the interweaving, the design!

The Warmth Organism of the Earth

Theodor Schwenk

If we want to develop a better understanding of the earth's warmth organism, we would do well to start by reminding ourselves of some of warmth's special qualities and of the way it interacts with matter. We know that the earth is made up of solid, liquid and gaseous elements, of land masses, bodies of water, and a mantle of air. Warmth is looked upon as a further fourth element. We rank it above the other three because of its unique ability to permeate them. In no other case can two bodies occupy the same space simultaneously. But warmth can do this. And in achieving this spatial permeation, warmth shows itself to a great extent free of spatial restrictions. Or we can put it another way and say that it makes its appearance at the borderline of the physical-spatial, now entering space, now withdrawing from it.

Warmth makes three kinds of connections with the physical world:

- *Conduction* is a phenomenon occurring mainly in connection with solid bodies.
- *Convection* takes place through the agency of air and water currents; they carry heat with them in the direction of their streaming.

Based on a lecture to anthroposophical physicians given at the Weleda, Inc. in Schwäbisch Gmünd, West Germany, in June 1974.

- *Radiation* is something we experience in nature at the hand of burning or glowing substances or when the sun shines on snow-covered surfaces. Indeed, snow is a very special case in point, for its heat-absorption and radiation reach almost the same radiation-maximum as do black objects.

The sun is the foremost source of heat, supplying 99.98% of the earth's total heat resources. When the air is dry, the temperature in the lower strata of the atmosphere falls approximately 1° C for every upward rise of one-hundred meters, while it rises by an average of 3° C for every one-hundred meter descent through the solid earth, except where there are special features like volcanoes.

Water plays a decisive role in the functioning of the earth's warmth organism for its thermal properties make it the great regulator of the climate and the weather. One thinks in this connection of water's huge capacity for storing warmth and of the anomaly of its achieving greatest density, not at the freezing point, but at 4° C, a fact of such far-reaching importance for every phase of life on earth that life could not exist were conditions otherwise.

Thanks to the detailed descriptions made by Rudolf Steiner on the physics of heat, its true nature can be read from processes in the atmosphere, where it is reflected.

For this atmosphere is indeed the very realm where all thermodynamic processes take place. We know them by names like isotherms, isobars, adiabatic changes,[1] isochors, and so on, and use these terms to denote changes that occur in the air when temperature, pressure, entropy, and volume remain constant. We have the most striking direct experience of the everchanging play of such warmth processes in meteorological happenings.

Patterns in Space

Now let us turn our attention for a moment to the spatial differentiation of the earth's organism into continents (lithosphere), seas (hydrosphere), and air mantle (atmosphere), noting that it applies equally to warmth.

At first glance, this spatial arrangement may seem purely accidental. But a closer look makes us realize that it is ordered. Quite aside from the fact that water and land masses are so distributed as to make for a water-hemisphere to the south and a land-hemisphere to the north, the shape and placement of the continents themselves cannot be looked upon as due to change. Rudolf Steiner described the continents to us as formed by the fixed stars, and he added that they float, as it were, on plastic-fluid bases—a statement borne out increasingly by recent research. This calls our attention to a fact often lost sight of: the tilting of the earth's axis in relation to the plane of its orbit, a constant through the changing seasons. Year-in and year-out it maintains its alignment with the north pole of the heavens as a still point. This can only be taken as evidence that the earth is "fixed-star" oriented.

This position, together with the earth's rotation around its axis, determines the way the continental land masses are distributed. And the possibility of their displacement, as a result of man's interference, is also being talked of these days—a displacement brought about by an irresponsible technology that is already effecting marked changes of climate all over the earth such as could lead to the melting of the polar ice caps and thus cause massive dislocations. Since these would take place at a rate of change so much faster than that measured by geological ages, they would lead to corresponding speedy changes in the orientation of the earth's axis, which would bring it into a different orientation to the fixed stars and the sun's path.

There is a further spatial patterning of the earth's warmth organism between the poles and the equator wherein we distinguish tropics, subtropics, temperate, and polar zones. This configuration is due to the spherical form of the earth and to the tilting of its axis. One can have an experience of the same patterning in a far smaller area, vertically rather than horizontally by, for instance, climbing 6,000 meter-high Mt. Kilimanjaro in Africa. The climatic zones through which one would pass in the 10,000 kilometer stretch from the equator to the poles are the same as those one traverses climbing to a height of only 6 kilometers. Here we see a striking heightening of vertical intensity over that of the horizontal between poles and equator.

The atmosphere in still higher zones changes to very low temperatures, which increase once more to a warmth layer of approximately 0° C at a height of about 50 kilometers. Then the temperature falls again to circa -80° C, only to rise with renewed vigor at yet higher levels. The uppermost layer (at circa 80–500 kilometers) is the earth's thermosphere. Thus, we find that the warmth organism of the earth is similar to that of all living creatures in that it possesses a variety of sheaths, including an outermost warmth-sheath within which life is able to develop.

The vertical and horizontal patterning encountered over the entire earth is repeated with a great many variations in smaller features of the landscape: in woods, fields and cropland, and rivers and lakes—all of which are separate warmth organisms and have their own individual warmth differentiations. Vegetation is the creator of these smaller warmth organisms. Areas where there is none turn into deserts subject to violent extremes of temperature, with burning days and bitterly cold nights.

Patterns in Time

After this brief glance at spatial aspects of the earth's warmth organism, let us turn our attention to warmth processes that run their course in time, for it is especially these that are bound up with our concept of all life.

All sorts of time patterns are to be found here—rhythms encompassing both the longest and the shortest time spans. Yearly and daily rhythms are the most familiar. We know them from the standpoint of many of their qualities, such as light and darkness, windiness and stillness, dry and wet periods, and rising and falling motions. But they are all warmth-related as well and may therefore also be termed "warmth-rhythms." They are meteorological and geophysical rhythms that are also often synchronized with astronomical rhythms. Since, as time patterns, they are a natural part of the same picture as spatial patternings, let us go on to study the interaction of these time and space aspects of the earth's warmth organism.

1. The "breathing of the continents." This refers to the great stationary high-pressure and low-pressure areas that coincide to a considerable degree with the continents and oceans and that rise and fall in seasonal rhythms. The well-known Asiatic High is a prototypal stationary high. In winter, it generates the Siberian cold

air masses brought over to Europe by sharp east winds. By summer, it has declined and become a pronounced low-pressure area, sucking up masses of moist air that cross the Himalayas and bring India its longed-for monsoon rains. European weather is determined by the stationary high over the Azores and by the Icelandic Low.

2. The arctic continents, consisting largely of ice, pulsate in tune with an annual rhythm: in March the sea-ice of Antarctica covers a surface of 2.5 million square kilometers that grows by September to an area of 19 million square kilometers, a 760% increase.

If we inquire into the cause of this breathing, this pulsation of the continents, we find it in seasonal rhythms that are themselves products of the earth's tilted axis. If it were not for this tilting, if, for example, the planet's axis were vertical to the plane of its orbit, there would be no such thing as changing seasons. The days, instead of varying in length, would all be of the same duration. Any change in the position of earth's axis would cause considerable instability in the yearly rhythms. The fact that it is fixed-star oriented and that the distribution of the seas and continents is fairly constant means that the rhythm of the year achieves that living balance that makes an "organism" of the yearly cycle.

3. An open mind can receive the impression that there is an underlying building plan behind these phenomena, which are in turn linked with further phenomena of a meteorological nature resembling organs and organic functions built into the earth's warmth organism. At special points in the sun's course, for instance, at its highest and lowest position, on St. John's Day and Christmas respectively, as also at the spring and

autumn equinoxes, air pressure exhibits tendencies known as "reflections" of its curves. Let us take as example the air pressure curve of a given locality a few weeks before St. John's Day and keep on observing it for the same period of weeks beyond that date. We often find that such curves reflect each other. The second curve might also be called "retrograde."[2]

Reflecting movements of this kind are well-known to musicians, for they are sometimes found in compositions of the great masters, such as Johann Sebastian Bach. There they provide a structural element, built up as though out of a higher system of laws into earthly time. Rudolf Steiner often called attention to the fact that spiritual laws work in a direction counter to the earthly time-stream. Structural concepts and laws of composition of a higher order than the earthly work into our world here, and we can discover and even measure them physically in annual rhythms such as those referred to.

4. What we thus encounter, particularly in annual heat and pressure rhythms, is met with again in the smaller compass of hourly and daily rhythms and much smaller spaces. I am referring here to something for which we might use a phrase borrowed from the yearly cycle, describing it as "the breathing of a landscape" in a single day's course. This breathing finds expression in a host of rhythms, such as the double rise and fall of atmospheric pressure, or in striking regular changes in the wind's force and direction along the seacoast or in mountain areas. We notice the wind coming from the valleys early in the day, from the peaks at evening, and speak of morning onshore and evening offshore breezes. Marked temperature changes often accompany these alternations.

Now what accounts for such rhythmic changes? Here again we come across large-scale causation: the rotation of the earth around its axis, which explains sunrise, high noon, and sunset; it also includes the whole range of thermodynamic phenomena in which we ourselves and all of nature participate, into which we are woven.[3]

5. Turning now to a survey of the various major and minor thermodynamic processes involved in the behavior of the weather, we will hardly be able to keep on speaking of anything as regular as a rhythm except in the sense of seeing it as the product of overarching rhythms of warmth and cold in summertime and winter. However, this very changeability in the weather is the basis upon which the constancy of the earth's warmth economy is built—a constancy essential to the preservation of all life on earth, every aspect of which depends upon it. Indeed, the living earth is itself involved in what happens in the weather, in that it is a biosphere affecting such meteorological processes as take place in the interrelationship of woodlands, cloud formation and precipitation.

6. Closer attention to what is happening in the weather discovers a law-abidingness there that is related to high-pressure and low-pressure areas passing above us. We all know about the old rules and sayings that go to make up the farmer's calendar, and we may have heard too of the "key-dates" that have been recognized from centuries of observation on the part of people whose work connects them more intimately with nature. Experienced meteorologists recognize them too, though they use different terms and speak of singular happenings during the year's course that occur in mete-

orological elements such as temperature, air pressure, moisture content, wind direction and velocity, and so on, all of which are linked in a living interplay. Many of these may be likened to processes taking place in organisms, and many a meteorologist schooled by outdoor weather observation draws such parallels. As an example, let us recall what P. Raethjen says in his *Dynamics of Cyclones:*

> . . . cyclones have a typical life history with characteristic beginning, developing and aging phases. They reproduce themselves . . . like a living creature in the sense that a young "frontal cyclone" is born out of the womb of an adult "central cyclone". . . . Cyclones have a metabolic process without which they could not exist, for they constantly draw fresh air masses into their vortices and excrete other masses in their outward spirallings. . . . We must not forget that this "dying and becoming" is an absolutely *basic* characteristic of cyclones. . . . Since the atmosphere behaves like a living creature, we see it truly only when we regard it and treat it as a totality. Phenomena which cannot be understood separately must be looked at in context as harmonizing aspects of a single whole.[4]

August Schmauss discovered in the year's course certain law-abiding principles, leading to the conclusion that the earth has a regulated warmth organism. He conceives the "dynamic year" as embracing the activity of meteorological elements: air pressure, temperature, and so on—shaping weather events in such an area as that between the western extremity of Brittany and the Shetland Islands to the north of Great Britain. This area is like a door through which Europe's weather blows in from the Atlantic Ocean. Schmauss uses the daily difference in air pressure between the two points as a reliable indicator. He takes the average of fifty years' findings on these dynamics of the atmosphere and comes up, contrary to all expectation, not with an irregular, haphazard line, but with a characteristic curve, which, as we recall, he interprets as follows:

> The "dynamic year," as we might call it, usually begins on the 29th of September with a bottom low along the pressure gradient from St. Mathieu to Lerwick and with a minimal atmospheric activity familiar to many as a phenomenon of Indian Summer. From this point on, the gradient—and the zonal circulation with it—increases in a series of waves until it reaches the winter high, falling on the average on January 9. The waves . . . are the expression of a battle between summer and winter, from which deep winter emerges victorious. But the culmination of its activity is also the turning point in the battle. Here winter begins its retreat with a series of rear-guard actions that show up in wave-forms in the meteorological curves. Activity decreases until it reaches the annual low-point, which occurs on the average sometime between May 22 and June 6. This stretch of days quite often brings the first heat wave, for there is not much movement in the air, and as the sun approaches its zenith it can therefore make its radiation fully felt.
>
> At the beginning of June this period of especially pleasant weather often ends quite suddenly. So-called summer monsoon weather starts—with heightening atmospheric activity. Cooler west winds pour in, frequently bringing with them along the wide front thunderstorms like those of India, the classic land of monsoons. The gradient remains high, though with wavelike weakenings, until September 16. Then there is a sudden plunge to the autumnal minimum, matching in pace the swift upward climb from the spring minimum. The calendar of events is thus by no means a matter of

> chance; instead, it actually deserves to be called a *collective timetable.*
>
> A meteorologist familiar with these facts sees in this scheduled taking-of-turns by masses of air something in the nature of an orchestral score that indicates where an instrument comes in—though on occasion it may miss its cue. Even so, it is a great pleasure to be familiar with the score. . .
>
> Our knowledge is quite equal to defining the type "low," but that is a collective term. Single lows are *individuals,* evidencing variations often brought about by very slight changes in the prevailing circumstances. We might, borrowing a musical term, call them variations on a theme. We marvel at the ingenuity of a Bach or a Reger who could write any number of such at will. Our atmosphere is certainly a master at this too; it need not surprise us that meteorologists marvel as they encounter ever fresh variations of a "low."
>
> Perspectives in the direction of *structural research* open up here; there is reason to believe that we are dealing with something very similar to *musical harmony in atmospheric events.*[5]

Here we have a meteorologist talking of musical form-elements in weather such as we also came upon as we familiarized ourselves with reflecting points in the year's unfolding.

We are led here, quite without our doing, to see a "design" in this "structure of the dynamic year," that of the pre-Christian and Christian festivals: Michaelmas, falling on September 29, the day on which the impulse toward a climbing dynamic, starting from the minimum, sets in; the Christmas festival of earlier times, celebrated on January 6 (in the dynamic year, January 9); Whitsuntide, from May 22 to June 6, a "movable" feast because it depends on the date of Easter—

and finally we see this same design in the reflecting points of the four chief seasonal festivals, which we have already talked about.

And now we find ourselves able to speak not merely of a design in time, but in space as well. The dynamic year "unfolds" in low-pressure or high-pressure vortices along the physical boundaries of the climatic zones, referred to at the beginning of this essay, *between* the wind or warmth zones. These are the so-called shear-zones, "sensitive" formations such as occur everywhere in nature and are built even into the organs of all living creatures. An example is man's inner ear, where the spiral pattern of the cochlea provides a sensitive receptor for musical and speech sounds. Formations of this kind are in every case zones where the slightest impact strikes upon "sensitive membranes" where cosmic forces can take hold. For the differentiated living being earth needs a corresponding world around it, as all living organisms do if they are to maintain their life. This world, for the earth, is the universe in its totality. The earth's "sensitive membranes" serve as entry ports for the forces of the sun and planets, which are able to project their ordering influences through them.

The tilted angle of the earth's axis gives it its fixed-star orientation. And this in turn makes the earth a sense organ with an unchanging focus in contrast to the ever changing play of sun and planets. We now understand what Schmauss means when he speaks of the possibility of lows that are due to make their appearance at a given moment in the dynamic year missing their cue and failing to enter. This can happen when certain planetary positions bear down too heavily on the dynamics of the year's course.

Now let us make a brief summary of the findings we have arrived at thus far:

> The stationary high-pressure and low-pressure areas are built up and dispersed again by the way land masses and oceans are distributed and by the climatic zones which form a belt around the earth with their lines of contact. They are, in last analysis, the product of the tilting and fixed-star orientation of the axis of the earth. The dynamic year is the outcome of processes taking place where climatic zones border upon one another in contact areas that are the product of the tilted position of the earth's axis, and are also the product of the path the earth takes in a year's course and of the influence of planetary movements. (A thorough-going study of the phenomenon of the "jet-streams" of the higher atmosphere would fully confirm and supplement this statement).[6]

Order in Space, Time, and Matter

It is clear from even so brief a summing up that the earth's warmth organism cannot be explored and characterized purely from an earthly standpoint, but must be looked at in closest possible conjunction with the solar system as a whole.

Now let us go a little more deeply into the matter of the tilted position of the earth's axis. We will take as a guideline for our own investigation a view held by Goethe and often fruitfully applied by scientists.[7]

No amount of effort enables us to succeed in fathoming the true nature of human beings. But if we observe their actual deeds, what they are "making happen," their character is immediately obvious.

Inquiring in this sense into the nature of earth's

warmth organism, into what happens as a result of the tilting of the earth's axis and of the earth's motion in relation to the sun, we find that just these simple facts alone are enough to demonstrate that we are dealing here with a dynamic filled with real wisdom, a dynamic without which none of these facts—including the distribution of the seas and continents—could have become a reality. A more thorough investigation of these facts will yield still more insights into processes of the earth's organism. We can feel it to be a pressing question whether this wise dynamic is not the underlying idea at work within the three factors already mentioned: the earth's tilted axis, its fixed-star orientation, and its rotation. We must ask whether there does not have to be a composer—the creator of the "score" and the time patterning—to account also for the spatial ordering through which the time pattern is realized.

The sun takes four steps as it travels its full yearly round at an angle vertical to its path through the zodiac:

- From the low-point on December 21 to the Equator on March 21 23°27′
- From the Equator on March 21 to the Tropic of Cancer on June 21 23°27′
- From the Tropic of Cancer on June 21 to the Equator on September 23 23°27′
- From the Equator on September 23 to the Tropic of Capricorn on December 21 23°27′

Put another way, every 90° advance along its orbit corresponds to a vertical step of 23°27′.

Thus, two components are involved in its motion whose relative lengths are approximately 4:1 (90:23.5–

3.84:1). Clearly, the 23°27′ tilt of the earth's axis has a threefold outcome:

1. A division of the year into four seasons;
2. A spatial division with a 4:1 or 1:4 ratio or relationship measured by the number of degrees in the angle;
3. A velocity scale in the same ratio of 4:1 as the ratio between the median orbital speed and the median speed of components involved in the ascending and descending phases of motion. In other words, the median orbital speed is approximately four times that of the annual vertical rise and fall.

So in the course of a year we can see a threefold configuration in time, space, and velocity, in which, according to the tilting of the earth's axis, the ratio of approximately 4:1 crops up three times.

May we take the fact that the earth oscillates in about the same basic 1:4 ratio as that governing man's rhythmic organization in the relationship of breathing to heartbeat as evidence that the earth is a living organism? Does there not seem to be a similar design underlying both, a design in the dynamic year's course that we also saw reflected in the Christian festivals celebrated at the principal points of the four seasons? We are obviously encountering here the same being that works as the ordering agency in time and space and that has been looked upon throughout the ages as the architect of the universe, the master of all harmonious proportion. For it is the degree of the angle of the earth's tilted axis that establishes the life-supporting relationship of earth and sun and has brought about the resulting arrangement of the seas and continents.

If these questions are justified and fit the case, the architectural laws involved must apply to every aspect large and small; this holds true for every living organism. Put another way, a fact of this kind should be ascertainable right down to the level of material composition, and we will offer evidence that this is so.

The speed at which sound waves pass through water is approximately 1400 m/s (meters per second), while the speed of their passage through the air is about 340 m/s. Here again we find the same 4:1 ratio. If we take into consideration the fact that the rate at which sound waves travel through a body depends upon its density, elasticity, and specific warmth, it becomes obvious that this rate of travel depends upon its material make-up and structure. Where, as in the case of the ratio between air and water, the 1:4 ratio turns up, it is an indication that this ratio also applies to the way the innermost aspects of matter are interrelated.

Thus, almost every property of water shows itself, in its relationship to warmth, as maximally suited to support the life of earth and its creatures. In this respect, water is unequaled, as L. J. Henderson has shown: the temperature anomaly, as a result of which water reaches its greatest density at 4°C and grows lighter again at the freezing point, is the reason why solid ice does not sink, but floats, thus keeping the earth from becoming a totally lifeless block of ice.[8]

Water's extremely great warmth-capacity is a chief factor in the origin of ocean currents as it is of all meteorological and biological processes. Warmth-capacity shows a dependence upon temperature as illustrated by the fact that at 37°C water requires the least amount of warmth to become one degree warmer. At the surface of the earth the sunlight is transformed into warmth. The water vapor and carbonic acid that are dissolved in the atmosphere possess a material composition that allows the sun's light rays to pass freely through them onto the earth, but keeps the warmth reflected from the earth from radiating back out into cosmic space. This characteristic has often occasioned the earth's atmosphere being likened to a "greenhouse" within whose protective sheathing (earth's air, vapor, and carbonic acid) life can flourish.

Interrelationships in Human Beings, Nature, and Cosmos

We can scarcely fail to notice that the human organism also embodies the same numerical ratio found, on the one hand, in cosmic and planetary interrelationships, and in those of substances in nature on the other. That this is a fact can be demonstrated in numerous examples. We will present some of them here, though they are not all taken from the thermodynamic field.

The 1:4 ratio so often met with in the design of the earth's organism reappears in human beings in the basic relationship of breath to heartbeat. We breathe about eighteen times a minute to the heart's seventy-two pulsations. As we contemplate the fact that man's airy and fluid organizations are founded on this pair of rhythms, it strikes us that the velocity of the sound-rhythm's passage through air and water is again approximately 1:4. This relationship is based on the innermost constitution of these life-substances (compressibility, density, and specific warmth). To take the case of water's specific warmth—in other words, of its warmth-capacity—we find that 37°C is its minimum—that this is the temperature at which water can most easily be heated. The correspondence here to the normal temperature of human blood—37°C—is obvious.

Focusing our attention for the moment on the temperature of 37°C itself, we find it occupying almost exactly the Golden Mean position on the temperature range between the freezing and the boiling points. (The Golden Mean makes its cut at 38% and 62% respectively of the distance between the two ends of a line or figure.) Can it be mere chance that the human form is likewise built on the Golden Mean; and that the architectural idea embodied in it manifests this ever-recurring proportion; and that man's body is composed very largely of water and in accordance with this law?

The human breathing rhythm is a reflection of a macrocosmic rhythm, that of the "Platonic Year," which spans 25,920 earth-years. For we breathe on the average 25,920 times in a day's course, and 25,920 is the number of days in a lifetime 72 years long. If we equate one day in the life of a human being with one indrawn and expelled breath, 25,920 such breaths will be breathed in a 72-year lifetime. Taking each earth-year as a single breath in the Platonic year, again 25,920 breaths will have been drawn in that period. The Platonic year covers the number of earth-years it takes for the vernal point to make a complete circuit of the zodiac. Thus, we may say that solar laws, too, are to be discovered in our breathing. The correspondence between our and the earth-organism's design and physiological processes permits us to see that both are based upon the same idea, right down into the very make-up of the materials of which they are composed. It is clear that human beings and earth are both products of a condensation process out of elemental warmth. People of earlier times were aware of this correspondence, and for that reason called the earth "Adam Cadmon."

Warmth as a Threshold Phenomenon

Since warmth always makes its appearance coupled with matter in a solid, fluid, or gaseous state, no independent being is ascribed to it. However, great quantities of heat are required to bring a solid body into the fluid or the gaseous state. This is simply to say that if a "body" is closer to pure warmth, it progresses further beyond the gaseous condition. This also means, of course, that its hardness and shape, and its solidity and structure, have disappeared and "gone up" in heat. Or we can just as well put the statement the other way around and say that a solid body is one that has gone through a condensing process out of heat; it passes through gaseous and fluid stages. This is a process observable in the evolution of spiral nebulae, fixed stars, and planets as stages of increasing condensation.

Thus, warmth comes to be recognized as having independent being and occupying a borderline position where the realm of the as yet immaterial—subject to spiritual ordering, structure, and ideas—passes over into the physically perceptible world.

The opposite obviously also holds true: Warmth, as the fourth element, stands at the transition point from the physical-material world over into the realm of ideas. And since human beings and world both bear the stamp of the same "structural concepts"—concepts of which we can develop awareness—the question naturally arises as to how the two relate to one another and what their effects on each other are including modern environmental problems. The relationship and the problems it gives rise to are so obvious as to require our developing an understanding of the earth's warmth organism. As we proceed to investigate human interaction with the world around us, it will be clear that the great environmental problems can really only be solved by reconnecting ourselves with the structural idea on which both the world and we ourselves are built—reconnecting ourselves, in other words, with that being from whom both once issued. This means acting in harmony with earth's building-plan. Then what we do in serving the earth will further our own potentiality for a healthy higher development.

Dis-ordering Tendencies in Technology

An extremely important thermodynamic concept, one that plays a decisive role in technology's dead world, is that of "entropy." Entropy is a measurable degree or level in a condition found in irreversible thermodynamic processes, but because it has to do with the relationship of energy to temperature it is not easy to imagine. According to the second law of thermodynamics every irreversible process is accompanied by an increase of entropy. When heat is used to generate electricity and to drive machinery, some heat is always lost; there is no such thing as 100% efficiency. Thus, an increase in "waste heat" is always taking place—a situation adequately described by the term "entropy." Practically speaking, almost every technological process is accompanied by an increase in entropy. As it goes on accumulating, it approaches a final value, a state of equilibrium no longer subject to change. All technological processes, then, run at an energy gradient, and entropy characterizes the reduction of an ordered state within a system to a less ordered one. Or, as Hagen puts it

> Entropy is the accumulating of a useless by-product made in the process of converting a system's energy or material components.[9]

We could also couch it in statistical terms and say that an increase in entropy results from progressing from an ordered to a less ordered state, from the improbable to the probable.

An example may serve to bring out the point of this discussion of entropy:

> Every irreversible process, left to itself, takes place in one direction only. Everyday experience demonstrates this. Molecules of a perfume dispersed in the air never go back of their own accord into the open perfume bottle. . . . They are always in favor of dividing up a big estate, but they will never willingly agree to accumulate a large estate for the benefit of a single favored individual, in this case the perfume bottle.[10]

Order-creating Forces in Human Beings and the Universe

Since nowadays such technological-scientific concepts as these are frequently imposed on the social structure, and every realm of life receives their imprint, the laws of breakdown, in the sense of ever-increasing entropy, have made themselves widely felt within the whole social structure of humanity. We see their life-destroying effects everywhere about us in the living environment.

And now, at this point, the opposite question immediately suggests itself: What happens when spiritual entities or human beings join forces to carry out some statistically unlikely action—an ethical one, perhaps, one in which there is a building up of order instead of breakdown? In such cases, laws of death, decomposition processes, structure-destroying functions are overcome. The opposite of entropy sets in, negative entropy, to be termed accordingly "negentropy." Is not a new kind of questioning indicated here—whether, for example, there exist realms in which lifeless laws like that of the preservation of power, energy, matter, the logical conclusion of which had to be the concept of entropy, are no longer fully valid? That such questions are being raised by biologists and researchers in the thermodynamic field is attested to by the presence, in recent professional literature, of reports like those of Riedl or R. Hagen. They discuss especially the work of H. Morowitz, Forester-Meadows and Glansdorf-Prigogine. The beginnings of this work date back to the 1940s, when Prigogine started concerning himself with thermodynamic questions about biological systems.

A picture very different from that obtained in the first half of the present century has come of this. Riedl says:

> The overall functioning of the biosphere promotes an increase of entropy, a quite extravagant one, at the cost of the flow into cosmic space of solar energy. However, local processes can create ordered patterns, such as are found in rotifers, in sonnets, or in Mona Lisa's smile. . . . One can call a living system a reservoir or storage place for negative entropy of wholly unimaginable proportions.[11]

When someone nowadays speaks in such connections of "the sun as the source of an in-streaming energy-flow *and* as the fountainhead of negative entropy," this must necessarily lead in consequence to new "revolutions in the physicists' conceiving of the world" and in the whole field of biology during the course of the next few decades.

Let us keep firmly in mind that "Entropy indicates the degree of disorder existing in a physical system." Where orderly patterns are built up, entropy decreases and energy is stored, as for example in biological systems on the most varied levels. But this means finding oneself in the realm where the laws of life obtain—and on sociological terrain.

The concept of negentropy as a standard of measurement for degrees of order has been fully accepted in thermodynamics,

> although for a long time it met with doubt and opposition. During the last decade, however, it finally crossed over into the field of biology, where its application is particularly important.[12]

Developments of this kind in modern science have yielded fruits of insight proffered by Steiner more than half a century ago as products of spiritual research. The "opposition" they met with among the scientists of that period was intense indeed.

Steiner spoke in numerous published lectures of a place where the doctrine of the preservation of power, energy, and matter—which gave rise to the concept of entropy—no longer applies, namely in man himself. He states that there matter is destroyed. Nutritive substances disappear and new matter is created, coming into being again out of the warmth that took them up, and this the more decisively the more enthusiastically a person cherishes moral ideals and an ethical ordering of life. . . . Here we have the essence of negentropy, expressed in somewhat different terms. For enthusiasm and active support of ideas, of ideals—in other words, of a creative "order"—are the opposite of decline and of entropy, which rules in the dead world of machinery.

So we may say that negentropy enters the picture wherever order is created out of chaos, as in creative thinking, and in human beings themselves. Of course our thought *content* is drawn from the world outside us. But in the way we connect and order thoughts we are not abiding by the dead laws of physics; we are free. Matter comes *freshly* into being in man's organism, in harmony with the formative forces of his patterning. These are characterized by Steiner as forces of buoyancy, not of a declining potential. The laws of the "sun-space" ("fountainhead of negentropy . . ."!) reign here, where, according to Steiner, there is no matter, only negative space, whence order-creating moral forces work out into the planetary system to make it an organism. And this is true also of earth's warmth organism as we experience it directly in the patterning and "score" of the dynamic year.

If human beings and earth are built on a pattern with a common basis, must there not also be places in the earth's warmth organism that harbor negative entropy? These would have to be areas governed by sun laws, since the sun is the source of negentropy. And the answer to this question is affirmative.

We have such places in the earth's warmth organism plainly in view during every thunderstorm. Recent investigation of the origin of lightning has uncovered the fact that there is a place within the earth realm ruled by sun conditions; there is even talk of the possibility of antimatter being generated there. We are told by the spiritual investigator that the highest hierarchies of spiritual beings are at work in the phenomenon of lightning. Here, then, is a realm of order-creating beings manifesting themselves in "cleansing and clarifying" thunderstorms. The findings of spiritual science and natural science could converge here. Even though modern thermodynamics has not yet included man in its research, the findings and concepts of thermodynamicists and biologists are already touching upon the secret of sun-space, described by Steiner as negative in contrast to terrestrial space and regarded today as a source not only of energy but also of negentropy as well.

Steiner does include human beings in the picture, speaking of the fact that while thoughts are being formed—in the process wherein the will rays from the warmth realm into that of thought—"lightning is really flashing on a microcosmic scale."[13]

There are three spheres in which human beings create negentropy:

- in knowledge created by will's activity in the thought realm,
- in art through the creating of ordered patterns in music, painting, and so on,
- in moral ideas possessing the power to create social order.

These realms are found to be those involved in the three areas of developing human culture: truth, beauty, and goodness, in other words, in science, art, and religion.

Negentropy in the biosphere: Ordering forces, called by anthroposophy formative forces, are at work in every plant, in every organism, even in the tiniest animal. In

Steiner's spiritual science, these forces are not merely named, each is exhaustively described in its unique characteristics and way of working. The earth's plant cover represents a realm, and a level, where negentropy reigns, and a process similar to the play of lightning in thunderstorms or to human thought-activity takes place when plants are pollinated.[14]

So negentropy, a living system of laws stemming from order-creating elements in space and time, is to be found

- in the warmth organism of the earth; in the year's course with its procession of Christian festivals and in the thermodynamic processes that take place in thunderstorms,
- in man: in the "striking in" of ordering ideas into the forming of thoughts out of the warmth element;
- in everything organic, and most especially in the plant cover and each single plant.

Perhaps we may quote a saying of Friedrich Schiller that draws its illumination from the realm we are presently discussing:

> If you seek the greatest and highest
> The plant can instruct you:
> What it is, will-less,
> Be that yourself with a will.
> That's the key![15]

It is clear from the foregoing that the force at work in what takes place in these three realms is the same order-creating sun-nature that is active in the orchestration of the dynamic year and in earth's warmth organism.

As we contemplated the design of the warmth organism described above, we inquired into the spiritual being that found expression in it. It is pertinent, in view of the breakdown taking place everywhere about us nowadays, to inquire into the nature of the causative agent here as well.

To find the answer to this question is also to discover solutions to problems of planetary extent that have arisen for the first time in the history of technological developments. For despite the tremendous contributions that technology has made to human culture, human beings have reached a point in their advances in it that permits them both to rule over nature and to plunge it into chaos and confusion. There is no question that far-reaching changes in the earth's warmth organism have been brought about that are already showing up in drought disasters and shiftings of climate. Air pollution, extending far up into the high atmosphere, has been increasingly cutting down the amount of sunlight reaching the earth. Vapor and combustion products emitted by jet planes have built up curtains of ice particles several meters high that are hanging in the upper atmosphere over the Atlantic and probably over other regions also; they continue building up every minute of the day and night. "Civilized" areas of every continent are covered by inconceivable amounts of smog that are being added to without let up. Rivers, lakes, and oceans are being similarly befouled. And here we are witnessing the start of a new episode in the human drama: the heating of bodies of water by the waste heat discharged from nuclear power plants.

The question was, what lies behind a technology based almost exclusively on an increase in entropy—governed, that is, by the laws of death? What *happens* in consequence in nature and the human race? Let us listen to the comments of some of the scientists on this point:

> If we apply this definition as an adequate standard of measurement for energy and entropy, the analogy to money suggests itself in relation to the concept of energy. In relation to the concept of entropy, characterized as it is by a disorganizing tendency to achieve a state of equilibrium no longer capable of change, the analogy to environmental pollution suggests itself.[16]
>
> As we know, our power to destroy negentropy in larger complexes has been put to use thus far only experimentally, but we are constantly destroying smaller entities. Oftentimes we become aware of the existence of some orderly pattern of relationships only when it has been annihilated. . . . Too dense and demanding population centers are the main cause of the chaos that is being wrought in such localities by overuse of energy. . . . What is lost to the biosphere is not energy—more can be generated—but the never-to-be restored larger environmental entities or biotopes that have been reduced to chaos. . . . And the growing use of energy is sweeping this order away at a constantly accelerating pace.[17]

We see that counter-forces are at work here on the warmth organism of the earth.

After what has been said thus far on the subject of earth's warmth organism, we can take it as almost symptomatic that technology and the economy interwoven with it have brought things to a point that can perhaps be described as a zeroing in on the sensitive spot in that organism. Geologists talk of climatic changes causing such a displacement of ice masses in the polar regions as a result of technological developments and overuse of energy that there is a distinct possibility of the earth's axis shifting in the foreseeable future. And if we take the warnings of responsible thermodynamicists and geologists quoted here as symptoms of the path that is being traveled by present-day humanity, we will have to say that the perspective Steiner saw ahead of us decades ago has unfortunately become actual fact, as he said it might when he was speaking about order being destroyed and civilization plunging into an abyss unless new spiritual impulses were taken up and put into practice, and science, technology, and business reconsidered their ways.

Man: Caretaker or Destroyer of the Environment

In providing access to the formative ideas underlying the design of man and earth, Steiner also laid the foundation for rescuing the warmth organism of the earth and of human beings themselves—foundations that are of vital importance for our time. For it has become possible for human beings not only to turn nature into a dead technological device but also to be so affected by the environment they have built up around themselves that they are transformed by it; and this is in line with the system of laws governing decline and the dissolution of created order. Having been given insight into the idea on which both humanity and the universe are founded, human beings are in a position to become the caretakers of the earth's organism. In Steiner's teaching about the formative life-forces he provided indications for a technology of order. For these forces are order-creating in the sense of negentropy, being related to the sun that is the latter's source. Steiner spoke of enthusiasm for moral ideals, which have their origin in our warmth organism and the soul—ideals related in their spiritual substance to the sun-space, for spatial separation does not apply in this realm of spiritual patterning. Steiner speaks on this theme in lectures given in December, 1920:

> Just think how our sense of responsibility is heightened by a realization that if there were no one on earth with a heart fired with enthusiasm for true morality, for spiritual ethics, we could not contribute to our world's continuance. This shining light which we generate on earth radiates effects out into the universe. We do not presently perceive how human morality rays out from the earth. . . . This radiance goes out for a certain distance only and is then reflected back again, so that we have here on earth the reflection of what man has thus rayed out. Initiates of every period have looked upon this reflection as the sun. For as I have often said, the sun is not a physical entity. What astronomy sees as a fiery ball of gas is simply the reflection of a spiritual element that makes a physical impression.[18]

With modern thermodynamics beginning to speak of the sun as the source of negentropy—in other words, as the source of order-creating forces—and with the question arising as to how thermodynamicists and biologists conceive a therapeutic approach to the present situation, let us hear what one of them is saying:

> The existing order has to be protected. For order builds on a foundation of order. And we must go on vigorously

> creating further order. And we are well acquainted with the cultural variants of negative entropy: humaneness and social order, science and justice, the arts and research.[19]

But where will we get the forces and knowledge with which to create this new order, since—in the last analysis—the situation now confronting us was brought about by applying the science thus far available to us? Is it not high time that we dealt seriously with facts learned from the science of spiritual ordering forces? If the sun is the source of negentropy, the final step in knowledge and action should be taken with courage and consequence: it is we ourselves who make our destiny. For the enthusiasm we engender in the pursuit of art, science, and religion is the same element that rays back to us reflected from the sun, the fountainhead from which humanity's and the earth's warmth organisms draw their nutriment.

Now that we have had an entropy-oriented technology for a century, we should turn our attention to the order-creating forces of negative entropy, of the sun's living system of laws, both out there in the universe and in our moral impulses.

There is reason indeed for enthusiasm and gratitude to Rudolf Steiner and his life work, which becomes of ever more decisive importance. And there is reason too to be grateful to many courageous modern scientists who have turned, in the face of opposition and skepticism, to new, creative ideas for the preservation of life upon the earth. We must hope that the consequences of their insight will be drawn by many others and that the concrete contributions of spiritual science to a new order will be understood and applied. Decisions of will rather than of knowledge are what is essential now, for the knowledge is already there. So it remains for every individual's free will to choose the forces with which he will ally himself,—those of entropy in the sense of decline or those of redemption in the sense of negative entropy, negentropy, and ascent.

Keeping the Earth the Place of Life

Theodor Schwenk

If we inquire into the foundations upon which all civilizations are based, we arrive at the perhaps curious but nevertheless definite fact that they are all tied to the presence of water. What would Egypt have been without its Nile? Every facet of Egyptian civilization depended upon this river as upon its lifeblood. What would have become of Mesopotamia without its Euphrates and Tigris, or of Greece without its springs and rivers, and coasts and islands? Primeval Chinese civilization owes its existence to its great river regions, whose sources are located in spurs of the Himalayas.

In earlier times, it was water that gave civilizations their character. Now, human beings imprint their character on the life-element water and destroy it, thereby destroying civilizations that have flourished for centuries. Is a civilization not a living growth, an expression of spirit-created order, of humanity's inner values, witnessed in art, science, and religion? We might quote here the words of one of humanity's great luminaries, Goethe:

> There is one justice upon earth:
> Each spirit to its own face gives birth.

Based on a lecture delivered at Herrischried, West Germany, in July 1975 to the supporting members and friends of the *Verein für Bewegungsforschung* (Association for Motion Research) and the *Institut für Strömungswissenschaften* (Institute for Flow Sciences).

For the inner imprints itself upon the outer and becomes the expressive countenance of our internal psychic-spiritual state. The same holds true of the water of our springs, brooks, lakes, and rivers.

Need this surprise us? How could it be otherwise? How are young people to interest themselves in a world pictured in "seats of learning" as nothing but a mechanism, even though all experience contradicts this?

What should our children report at school of their experience of the weather if it is just a machine, a piston-powered engine? Or say of the earth as a living realm if it is only a speck of dust in the universe, bare of spiritual aspects, lacking all ordering? What are we to think if there is no such thing as a higher development, if the human being is a mere animal, the chance product of a combination of molecules, or if, as a well-known Nobel prizewinner recently put it, "Life is (the product of) chemistry." Need we be surprised that many people brought up with the phrase "the battle for existence" being sounded in their ears have drawn its consequences in their behavior?

We would not have to keep on reviewing these familiar matters if they did not point to a process in which we are all more or less involved.

Two courses run parallel: the collapse of human civilization and the concomitant falling apart of the living organism of the earth. But the latter reacts again upon the former, since youth cannot find anything to look up to that makes going on living seem worthwhile. The consequences of the kind of physics and chemistry described here are realized in the social sphere as something in the nature of a chain reaction, with a dissolution of order in both human society and in nature. A process akin to a tailspin is going on, in which interdependent civilization and nature alternately affect each other instead of being shaped by men in freedom. How are we to extricate ourselves from such a vortex? It is plainly a case of the difficult task described by Friedrich Schiller in his *On the Aesthetic Education of Man:* of understanding this interplay between civilization and nature, and of restoring it to order while it is in motion.[1]

We must certainly ask at this point whether science has shown concern over the direction in which it is being propelled as a result of its own doing, and what conclusions such a retrospect has reached.

Let us turn for an answer to a work written by a man who has set himself the task of studying the deeper reasons for the dislocations of balance that are occurring in natural processes all over the earth.

We refer to the book by Nigel Calder entitled *The Weather Machine.*[2] Calder discussed weather and climate with leading authorities in a dozen countries. He has published his findings, concluding that there is a whole series of reasons for fearing that a decisive worsening of the earth's climate is in store for us. He reports on a number of indications pointing in that direction. Some of these were catastrophic drought in India and countries bordering on the Sahara, crop failures in lands where good harvests were formerly the rule, the advance of the Arctic and the concomitant shifting to the south of deserts in the Sahel, where, at the time of the book's appearance, ten million people were fleeing before the drought. Calder describes his impressions, obtained at highly specialized institutions such as the Geophysical Fluid Dynamics Laboratory at Princeton, where 100 calculations dealing with numerical data on climate can be carried out in one-millionth of a second. A model there represents the atmosphere at 18 different levels. It can be used to elucidate past changes in climate and to predict future changes. The Princeton facility for the study of climate is known to be the most capable in existence, but it too is incapable of meeting the need. This can give us some idea of the inconceivable multiplicity of the factors affecting our planet's weather and climate, bringing about interaction of a complexity encountered wherever life-processes take place.

This raises the question whether it is at all possible to deal statistically with seemingly minimal influences on the weather that affect large energy accumulations in the atmosphere and, under labile conditions, are capable of bringing about immeasurable consequences.

Calder encountered the same skepticism in other investigators. Despite the valuable insights into the physics of weather and climate that computers provide, these authorities had become convinced that

> There are probably too many complexities and indeterminacies of detail for the numerical model ever to produce the actual forecast for the periods concerned in seasons, years, and decades ahead.[3]

These are the words of a particularly noted climatologist whose state support was cut off at the behest of the meteorological establishment because he dared to state an unwelcome truth. We report this fact because it shows what can happen when, as a result of the intertwining of the interests of the scholarly world with political and economic interests, the disclosure of certain truths is not permitted. Natural occurrences could bring them to light under some conditions, but that would be too late.

In the instance cited, the theoreticians rather than the empiricists won out, although it was the empiricists who gathered the experiential data fed into the computers.

Calder speaks incredibly of a "weather machine" and gives his book that title, even though he admits that weather and climate are such complex systems and so inextricably bound up with each other's gyrations that no one can distinguish between cause and effect; they baffle the most modern computers and make nonsense of any talk of control, the distinguishing mark of a machine.

We will return later to a discussion of many of the insights afforded by Nigel Calder's work. But for the moment let us concern ourselves with another picture of our planet, the picture of a living organism.

The Earth as Living Organism

The concept of the freeing of latent energy first formulated by Julius Robert Mayer and later elaborated by such meteorologists as Albert Schmauss enables us to do away with the myth that the energies put to use in human technology are much too small to have any effect on weather and climate.[4] Since modern humanity is far too poorly grounded in an understanding of the freeing of latent energy, the lack of it produces a harvest of weighty consequences. We underscore our incomplete understanding by citing the words of a leading meteorologist whom Calder also quotes in his book.

> He [Edward Lorenz] emphasizes a peculiar characteristic of the weather and its mathematical descriptions, that an effect can be as disproportionate to its cause as a forest fire started by a cigar butt. In weather forecasting

> it means that a very slight difference in the prevailing weather—the presence or absence of a thunderstorm for example—could make a substantial difference to the world's weather a few days hence. . . . But a thunderstorm can double its intensity in 20 minutes. . . . In practice thunderstorms are too small to register in the network of global stations, or to figure in the computer's calculations. But they typify the subliminal sources of error that can eventually frustrate attempts at long-range numerical weather forecasts.[5]

Is it not characteristic of the living that it carries on essential functions with the least expenditure of forces, energy, or matter? Is there not a similarity, in such cases, with the familiar picture of a little boy leading a bull to pasture that is simply bristling with power? It is always trace elements in the living that decide the issue, right up to and including weather happenings, with the factors that set them off, such as condensation-nuclei in the atmosphere, and right up to and including maximal eccentricities in planetary orbits within the ordered organism of the cosmos. They are decisive not only in setting off labile conditions, but are often even capable of bringing such conditions into being.

In previous lectures—"Water: Destiny of the Human Race," for example—we called attention to facts of this kind and to physiological processes in the earth's organism. We remind our listeners very briefly of what was said on those occasions about the process occurring in all the world's oceans, which causes waves to travel from regions of bad storms, the short waves remaining behind the long waves as the latter travel thousands of kilometers to show up, days later, on distant coasts. As we reported at that time, corresponding processes take place in the fluid-filled "snail" or cochlea of the inner ear; the long "waves" of deep tones traveling through it to the very end, whereas the shorter waves are eliminated immediately upon entering the labyrinth. We stated too that the process in external nature can be viewed from the human angle, looking upon it as an extension into the environment of physiological processes found in man.

We pointed also to the system of eddies or vortices in the North Sea, operating there like a heart organ in the midst of the sea. And we called attention as well to the material components of sea water, the composition of which has been found to be analogous to that of human blood, even in the matter of quantitative proportions; the only exception being that where human blood contains iron, ocean water contains magnesium. This characteristic indicates a relationship between sea water and plant life, in that the world of green plants has built magnesium so predominantly into its fluid organism, pointing us again to the realm of life. On every hand we find evidence of functions typical of life and of the human organism expanded into the macro-organism of our planet, making it seem the more surprising that Nigel Calder speaks here of a machine:

> Around the equator, where the cloud clusters pour out their moisture on the rain forests of the Amazon and Congo, there are deserts or near-deserts. For example, air that rises over Indonesia sinks over the parched Galapagos Islands on the other side of the Pacific. . . . Changes in the positions of rising air and sinking air around the equator, or intensification of the movements, are responsible for unexpected droughts and floods. . . . Associated with these shifts are changes in the ocean

> currents but also changes in weather and weather patterns in the stormy zone. Which is the cause and which the effect among so many, widely scattered parts of the machine?[6]

We look, in short, through such facts and functions as through a window opening on a human being extended in space, on a physiology akin to that of a living being—clearly the model for the earth's structure. We spoke of the cosmic image of man inscribed into the earth as of a design brought to realization and maintained by rhythms. Calder too speaks repeatedly of rhythms in his summing up of the impressions he received on his journey around the world to confer with its leading climatologists. We heard of this before on the occasion of reporting research done by Albert Schmauss, where he speaks of musical patterns that find expression in the annual dynamic cycle of the atmosphere, the high points of which tend to occur around the times when the major Christian festivals are celebrated, festivals to which nature thus makes her contribution. Schmauss speaks of a score perceptible in the year's course, with passages reminiscent, in their themes and variations, of a Bach or Reger—in other words, rhythms upon rhythms, symmetries, heights and depths such as are characteristic of the great musical compositions.

We naturally ask who is playing this score.

What is the starting point for obtaining an answer to this question, one that presents itself as a matter of course?

Let us state it without further ado: the main starting point of all the rhythms involved in these living processes is to be found in the tilt of the earth's axis in relation to its horizontal orbit. This is what causes the seasonal pendulum swing of the sun as it climbs toward summer and descends toward winter; there are four steps during each revolution of the earth around the sun:

- From the end of March to the end of June, an ascent to the zenith;
- From the end of June to the end of September, a turnabout and descent to autumn;
- From the end of September to the end of December, a further descent to deepest winter;
- From the end of December to the end of March, a turnabout and ascent to spring.

There are thus four "vertical" steps to a single more or less horizontal revolution. Thus, the carrying out of the year's time-schedule shows a ratio of 1:4 in relation to the rhythm of the four seasons.

It should not be difficult to figure out what changes would result if the earth's axis were to assume a vertical relationship to its horizontal plane of orbit. The answer is that there would be no seasons, only times of day, and only polar regions and tropics, with a relatively narrow belt between them. The polar regions would greatly predominate and any life that might still be possible would be hemmed into that thin strip. The graph of temperature in every part of the earth would be a straight line rather than a rhythmical curve. If we wanted a change, we would have to make it for ourselves by moving from north to south, or vice versa.

But if the axis were to lie in the horizontal orbit itself and remain parallel to itself there during the year's course, as it presently does, there would still be many possible directions of the axis within this horizontal orbit. One thing would be common to all, however: there would be seasons, but they would pass so much more swiftly that every part of the earth would be exposed to the sun's rays in such a way as to undergo alternate exposure to polar and tropical conditions. The sun would not only climb in a period of three months to the tropics of Cancer and Capricorn, but would reach the Pole. In other words, all four seasons would experience an intensity and tempo of fantastic proportions, and the annual temperature curve would oscillate from one life-negating extreme to the other. Life would not be a dull monotony, as in the former case, but a fantastic existence.

The fact that the earth's axis maintains its own prescribed tilt makes it possible for the earth to have its three familiar climatic zones. This makes for a middle zone of considerable breadth, characterized by the rhythms we were justified in describing as musical. Space is created for a temperate middle zone, a scene of equilibrium, as well as of motion, where the four yearly seasons come into being. It is a space where balance can exist, or—put in a way that means the same for humanity—where culture can develop.

The spatial measurement of the degree of tilt in the earth's axis is known; it is 23.5 degrees of arc, which conditions the position of the tropics in the heavens and on the earth. But 23.5 degrees is approximately one-quarter of the angle between the equator and the poles, which confronts us again with a 1:4 ratio, this time in relation to space. This finding should be kept in view during the course of our further study: from the aspect of both space and time, the earth's organism is built on a ratio of 1:4.

Are we justified in speaking of chance here, when the identical ratio turns up twice in the design of the planet earth (comparable to that of a functional human being on a cosmic scale), and is found again in every one of us as life's basic rhythm? We would rather express it thus: If the direction of our thinking as it takes account of these facts is valid, then the following must be true: the basic rhythm of human life, the relationship of pulse and breathing, brings the same law into operation with 18 breaths to 72 heartbeats in a ratio of 1:4.

Ernst Bindel writes in his *Numerical Basis of Music* that

> . . . the ego that has its being in our blood is the conductor who decrees this 1:4 tempo for our music-making life-spirits as we play the symphony of our lives.

There is increasing confirmation that the earth and human beings are built on the selfsame pattern. And the question arises, who, in the case of water, the life-blood of the earth, is the conductor imparting this rhythm to the life-spirits there as its life-symphony is played? It must be the same conductor who, in times when lofty culture reigned, was called the master builder of the world, appearing in works of art with a pair of compasses and square, working as a geometrician arranging the world order and keeping it timewise and spatially in balance.

Let us listen again to Nigel Calder:

> There is reassurance in the fact that, despite enormous changes in geography and in the composition of the air, our planet has so far avoided the two extreme conditions and has merely varied between Earth 2, Earth 3 and Earth 4 (overheated and chilled). There seem to be profound natural regulators at work which maintain the tropical temperatures, at least, within fairly narrow limits and keep the all-important composition of the sea water more or less constant. But humans will be well advised to learn about these global regulators and make sure that they do not unwittingly override them.[7]

In view of the current situation this can only mean that humanity has got to acquaint itself with the master builder of the world and learn to understand his work: the earth as well as human beings themselves. We see this as the way to get an answer to our question as to who the conductor is.

Can we find out anything from the meteorologists and climatologists to whom Calder refers that points more concretely to these regulators, bearing in mind that the concept of regulation cannot be entertained apart from the concept of spatial and temporal planning? We can answer this question, for Calder speaks in another connection of the fact that the tilt of the earth's axis is, strangely enough, the only constant in a picture where change has been the order of the day for centuries and millennia, having varied only between 21.8 and 24.4 degrees. In other words, changes in the tilt of the earth's axis keep to the 1:4 ratio, shifting, like life itself, first to one side, then to the other, with very little effect on this ratio.

Here, we believe, is that wise regulator we have been looking for, the magic baton of the conductor who adjusts the angle to the whole surrounding cosmos and thus makes it possible for the dynamic year's course to proceed in rhythm. A most vital life-regulating principle is clearly to be seen here. The axis, however, is not a material reality. And yet everything revolves around it! May we not regard it as the secret of the great bodyings-forth of the cosmic design that their most vital characteristic is not carried through to the point of material embodiment, but is allowed instead to remain invisibly present, resounding through space like a living being? Nor is it by chance the secret of the great cathedrals that their proportions conform to cosmic laws, to work the more powerfully through their invisibility upon the "shaping" of people who spend time in them.

Though Calder is unaware of the deeper reasons for it, he finds himself impelled to call the earth "the best of all worlds" in contemplation of the unique regulation described above. And L. J. Henderson, who has made a thoroughgoing study of the physical regulations governing living organisms, comes to the conclusion that "The present environment of the life-forms of the earth is the best suited to its every form that we can conceive. . . ."[8] Henderson also sees the unvarying composition of sea water referred to above as the basic factor enabling man to develop civilizations on earth as the cradle of life; and that constancy, with its life-supporting capacity, is again the result of the tilted position of the earth's axis.

Let us now sum up in a series of key sentences what the meteorological and oceanographic research linked to the names of such investigators as Schmauss, Henderson, and Calder whom we have quoted has brought forth in the way of vital insights:

1. The earth organism founds its various functions upon a multiplicity of rhythms.
2. These are optimal, in the way they are patterned, for the needs of life.
3. The patterns upon which the functions of the earth are based are of a musical nature, and therefore necessarily imply a time-structure. They show up at the festival seasons of the Christian calendar.
4. They are the basis of a regulative principle with which the humanity of our time must of necessity acquaint itself if the earth-organism and

with it human beings themselves are not to suffer serious harm.

5. The temporal patterns are built on a ratio of 1:4.
6. These patterns are bound up in turn with the spatial pattern of the tilted position of the earth's axis, which also exhibits the ratio 1:4 in its right angle.
7. This brings about optimal life-conditions, so that the earth must be recognized as "the best of all possible worlds" for the development of life and of culture.
8. The rhythms of the earth as a macrocosmic organism are reflected in the structure of the microcosmic human organism, most notably in the basic 1:4 rhythm of heartbeat and breathing.
9. The question arises: Who is the conductor responsible for these spatial and temporal patternings?
10. The answer is to be found in the fact that a meaningful whole results from the working of the factors mentioned. This justifies us in speaking of a living being active in these patterns. The "regulative principle" is indeed just such a being, and it is time that humanity recognize it as such to avoid harming both itself and its dwelling place.
11. The earth, then, most certainly possesses no such thing as a "weather machine" or piston motor in its atmosphere. It is not just a grain of dust in the universe, made up of haphazardly assembled molecules. It must rather be likened in its functions to a musical instrument on which the master builder plays cosmic harmonies. This makes it fit to bring forth cultures

and offers a unique opportunity for a schooling of human spirits, for the development of individuals whose goal it is to become freely creative beings.

The Future of the Earth

Once the building plan has become thus manifest, all further details of nature derive their meaning from it. The further course of today's inquiry into how the earth can be kept a living realm is self-evident; it calls for dealing with the following questions:

1. How should the unhealthy relationships between our actions and the earth-organism as a whole that are still in effect today be reshaped in future? An accurate diagnosis will indicate the cure.
2. What is the positive course to be followed
   a. in insight and action
   b. by individuals as well as by society?
3. Have human beings already attained the capacity to manipulate life-forces?

As to item 1., the relationship of the small to the large: It is always being asserted that huge natural catastrophes have occurred over and over again in the past and that human beings are much too small to do the earth as a whole any real harm, in the sense of upsetting the balance of its functions. The facts show views of this kind to be outmoded. Thoroughly responsible professionals point to the fact that, considering the human assault upon the natural order presently occurring on an unprecedented, global scale, possible dislocations of the earth's axis cannot be ruled out. We need only picture how changes in climate brought about by people can cause massive shifts in polar ice zones that would bring the angle of the earth's axis out of adjustment. As we have seen, the controlling forces need not be tremendous ones; nature habitually works with small measures.

So our question should be: Where are they to be found in the human being whose acts affect the natural order? Where are they to be found in us?

In our organic functions the most delicate formative forces are the energy-organizers; they are actually spirit-created patterns, or at least should be. But they are affected by the drives inherent in human soul-life. If, through lack of insight, these are not subordinated to the laws of life, illnesses and crises follow. One may, for example, be impelled by such drives as "the battle for survival," by ambition, or the lust for power, or by a desire to manage everything, with all their attendant hectic consequences. The outcome is heart and circulatory trouble, the number one illness of our time. The ratio of 1:4 is upset; arrhythmia develops. Here we have the prerequisites for the onset of bodily and psychic illness.

What is really at the bottom of drives like the above? They are the product of a mistaken picturing of human nature that obscures the plan on which man is built, insidiously replacing it with a view of him as the highest animal. Must not the further consequences of such misconceptions be the proliferating of psychiatries? The therapy is the offering of a new picture of man and of his "brother," the earth, his cosmic counterpart that has long been available in the anthroposophy of Rudolf Steiner.

Considering the correspondence that exists between

microcosmic and macrocosmic man and the ruthless global-scale disruption on which humanity is now embarked, we find these imponderable psychic driving forces are manifestly the agents and causes of the present-day problems of the earth and humanity. There is ample proof, even though in the form of negative consequences, that imponderable psychic factors are capable of fundamentally changing the earth. We need only think of the unrestrained exploiting of the earth's mineral wealth being carried on by interests with no concern for the total terrestrial-human organism—interests willing to accept irreparable damage in the form of desert wastes as "the price of progress."

As to item 2., the positive course to be followed:

When scientists at the Massachusetts Institute of Technology speak of "a revolutionary change of direction of Copernican magnitude such as has never before been demanded of the human race," they are accurately stating the situation now confronting us all over our planet. But the question is not only what the revolutionary change of direction ought to be but how it is to be accomplished.

Since, in the last analysis, control is an attribute of the inner life of the individual and takes effect on the larger, outer scene from within human beings, so that from now on the control of the earth-organism is increasingly placed in the hands of human beings, the questions raised above must also be answered from within them:

- Through insight into the being from whom the earth-organism and the design of man himself issued,
- Through action undertaken in harmony with that insight,
- Through putting a stop to the greed for power

> and money and position, and ceasing to live arrhythmical, hectic lives—in other words, through shaping civilization in ways consonant with the dignity of man, viewed in the light of the cosmic thoughts by which the world was shaped.

It might be objected that the earth's inhabitants have fostered this attunement to the world-creator in every age, and never more so than in Christian rituals. The world has undoubtedly been the recipient of an unending stream of blessings from such sources. The situation is nevertheless as described in the quotations with which this talk began.

Why is this the case?

Because the time-spirit of the earth demands a breakthrough to the spirit on the part of science. Modern consciousness has evolved, through its exposure to science, to the point where humanity wants to learn to understand and have actual experience of what used to be accessible to belief alone. What is needed, in addition to a spiritually-minded science, is a genuine science of the spirit at work in man and the universe.

Challenges of this kind easily pale to mere slogans unless they are read from immediate experience of the time's dire needs. Words of Rudolf Steiner lend depth and clarity to what is meant here:

> Christianity will not have been properly grasped until it has permeated the earth right down into the physical sciences. It will not have been understood until we grasp, right down to the physical level, how Christian substantiality works throughout the entire life of the universe.[9]

The time has come for human beings to ally themselves with the spirit who, as the master builder of the cosmos, united itself with the earth and "will come again in the clouds," and to take upon themselves, through their own actions, the responsibility of the further control of the building plan of the earth's organism. The times demand that "revolution of Copernican proportions" in which every individual can take part in full awareness of the total picture.

Let us briefly review some aspects of a life-oriented technology taken from previous lectures and from our considerations today.

A basic reorientation of technology will have to take into account cycles inherent in the living element. In our present striving for 100% efficiency we interfere with life cycles, with a consequent diminishing of life.

This in turn involves the necessity of building up structures that conform to organic functioning, whereas, in the case of the technology of mechanical forces, energy can only be derived from a breaking down of organic structures.

We must therefore see to it that a balance is maintained between tearing-down and upbuilding processes in technological matters if further damage is not to be done by one-sided destruction.

To achieve this, every living organism must be coupled with an environment with which it mutually interacts. A scrutiny of industrial areas will show that just the opposite is taking place; such interrelationships are being wiped out everywhere. It is no accident that we use the term "industrial deserts."

Concrete beginnings of cooperation between men and machines are now apparent in deliberately undertaken adjustments leading to more rational practices, such as, for example, the abandoning of nonstop assembly lines.

Adjustments of this kind are always attuned to the lawfulness involved in mutual relationships between not just two, but many components, so that causes and effects are linked in organic cycles.

The technology of the future will have to give special consideration to the cosmic rhythms in the fluid element rather than to a rigid mechanics in which all numerical relationships fit neatly into rigid numerical proportions.

Nature always works, as we have seen, with small deviations from the norm, with numbers that leave remainders—numbers that mathematicians call incommensurables, which condition all cosmic processes and keep them in motion. Incommensurables represent the "open space" that always keeps a process from becoming mechanical. Often these are tiny eccentricities such as we have repeatedly referred to as being characteristic of organic laws.

The recent striving of our innermost being has been to find the Tree of Life, to add to our possession of the Tree of Knowledge. All our ponderings and aspirations can be so described. Of course, everyone still wants his or her own particular Utopia as well.

Engineers and the constructors of the new biotechnology will also have to familiarize themselves with the laws governing the life-realm, in order to work in harmony with living systems. Those concerned with efficiency will have to take into account the time-factor involved in living evolution. To neglect this, to work with linear rather than with organic time will, in the long run, mean lessened efficiency. In witness whereof we see the humanity of today paying the bill for the peak economic efficiency of yesteryear.

Our second question was, What is the positive course to be followed, in insight and action, by the individual, and by society as a whole?

It will become increasingly a matter of course for everyone to view the living organism of the earth as the body of a spiritual being, like many lines of force moving in the same direction and producing a force-field that, as a higher entity, receives its life from above. Though this resembles what can be experienced in a scientific setting, it will come about as a result of the free decision of each individual. When every plant responds to an inner urge to turn toward the light, a field of living forces comes into being, like a field of ripened grain in ear. Friedrich Schiller expressed this when he said that man must carry out consciously what the plant does involuntarily. Every one of us can keep in mind and serve life-principles, so that a balance is brought about between a technology of the lifeless and a technology of the life-realm.

This synthesis accords with a story-symbol coming down to us from ancient times that pictures such a development: When its inhabitants were being ejected from paradise, Seth begged to be allowed one backward glance. The archangel granted this request, and Seth, looking, saw that the Tree of Knowledge, of whose fruit the first human beings ate, had grown together with the Tree of Life, forming one tree.

This picture must become a reality today for the human race to achieve its longed-for goal. The technology of the physical realm, a product of the Tree of Knowledge, will have to merge and form a synthesis with that other as yet neglected half of the world, the cosmic Tree of Life. The only question is that listed above as the third of three, namely, Have we progressed far enough to achieve this without doing further harm?

We are still like sorcerer's apprentices who would upset all balance everywhere if we were to put life-forces as such to work in technology. We have failed the test of many an experiment in this direction. An example is the manipulation of the weather referred to earlier in this talk. Floods from the fluid sea of life would be set in motion, as Goethe pictured it in his poem, "The Sorcerer's Apprentice." For what sort of forces are these?

They are of the world of cosmic water-forces, of the sea of life-forces itself as this pulses in the numerical relationships of the music of the spheres. They are of the world that directs the shaping, formative forces, organizing, and patterning organic forms.

That is why a strict path of schooling has to be traveled toward this goal, toward the achieving of the magician's mastery. It involves

- reverence for everything alive, for living patterns,
- sacrificial transforming of egotistical drives, with constant attention to the whole,
- connecting ourselves with the life-spirit of the earth,
- founding schools of higher learning that would not merely teach competence in understanding the lifeless realm, but would also inculcate an enlivened thinking such as is alone capable of opening the door into the world of life that has its being in the weaving dynamics inherent in living processes.

We pointed out that

- there can be no change without a schooling in reverence, and without change no connection with the Tree of Life.

These will in turn foster an enhanced capacity for reverence, profounder change, greater harmony with the true, the good, and the beautiful—in short, an ascending spiral as opposed to the descent that has led to so much misery and dismay.

We will not gain possession of the secret of life or of the key to it by a furious assault upon any such goal, but only in upward-moving cycles. We are told of this in traditions that have come down to us from ancient times in the form of myths, fairy tales, and sagas. They invariably refer to a treasure of some kind, and the treasure is always hidden under water. And the question arises: Is this treasure not the secret of water, the secret of life itself?

There are legends in which the treasure is a secret that involves a ring:

The Tree of Life is growing on a high mountaintop. There is a lake there, with a white swan swimming on it. He is the guardian and bearer of the secret of water, for this secret has to do with the golden ring which he carries in his beak. If any mortal touches the ring or gains knowledge of the secret without undergoing cleansing, all the sluices of the ocean of heaven open and destroy the world, which would also be brought to an end if the swan were to drop the ring, which the Lord God Himself put into his beak to keep the world in balance.

Legends of this kind, collected in the preceding century, are like prototypal pictures of what is about to happen to the human race today, the situation described by Nigel Calder too, in warning, as the basic message of his book:

> If the trend also means persistent drought in the tropics, mankind may be in very deep trouble indeed, never mind the Little Ice Age or the Big Freeze. . . . Perhaps we shall be lucky. In the nick of time we may comprehend the problem and cope sensibly and humanely with unavoidable disasters. . . . We may even bury our political differences while we take action to arrest a dangerous change in the weather. But my optimism flags when I think of one man's shipwreck being another man's harvest, and of climatologically well-informed nations gaining money or strategic advantage from other nations' disasters.[10]

And we add: The tailspin of culture, of civilization, can be reversed if we add the spiritual-scientific impulse to the natural scientific impulse living in the human race today. To the practical view there is no other alternative if the earth is to be kept a realm of culture, and thus a realm of life as well.

The Human Role in the Earth's New Connection with the Cosmos

Theodor Schwenk

If we are to deal effectively with the water situation that is becoming such a serious worldwide problem we will have to proceed on the basis of an insight into the essential character of water, that spiritual being whom we can come to know in nature by observing its "deeds" and attributes. No resurgence of reverence for the life element can come about without our developing such insight, and a failure to do so would leave us without a therapy for treating the causes of the current situation. Steps in the direction of attaining the necessary understanding were suggested in my previous lectures.

The last one, too, dealt with this aspect—with its inquiry into how the earth can be kept a realm of life. In view of both the declining quality and quantity of the water at our disposal, this question is extremely urgent.

Since we have already tried to convey a picture of the great wisdom with which the earth and its life-functions were designed and ordered, we can go on today to consider the way its organism is interwoven with that of the great cosmos. Our role in this will, of course, have to be included in the discussion.

Our theme, "The Human Role in the Earth's New Connection with the Cosmos," at once raises the question,

---

Digest of a lecture delivered at Herrischried, West Germany, in the summer of 1977 for friends and supporters of the *Institut für Strömungswissenschaften,* the Institute for Flow Sciences.

"Why is the connection new? Wasn't there always a connection, without our playing any part in it?"

Previous lectures dealt with the theme of the noticeable disturbances that have already taken place in the life of the earth in relation to the cosmos. We had to call attention to climatic changes and to the massive dislocations of balance that could result from an increased icing up of the polar regions. We also pointed to possible related shiftings of the earth's axis, bringing about ever more drastic climatic upsets, in short, to instability in the patterned order over large areas of our planet. Whatever views may be held on the subject of this instability—including that of a possible melting of polar ice masses due to a heating up of the atmosphere in consequence of air pollution—it is clear from the increasing industrialization, the exploiting of natural resources, and the unwholesome agricultural practices being engaged in all over the world, that human beings are invading dimensions where they are capable of disrupting and perhaps even of destroying planet-wide patterns. But it is also clear that from now on we must provide conscious, active support to the functional connections between the earth and the cosmos. The picture of things presented in these talks is intended to call attention, over and over again, to such overarching organic relationships and to sources of insight on the basis of which practical supportive initiatives can be undertaken.

Let us, to begin with, review the relationship of man, earth, and the cosmos that existed as nature created it before the advent of our age of invention and technology.

Humanity was still completely interwoven with nature and with daily and seasonal rhythms conditioned by climatic zones. Whether we contemplate the lofty cultures of the Orient, as far east as Asia and Siberia, or the Andean populations of South America, they all had one thing in common, namely, an intense experience of the way natural forces permeate and rule the world as living beings. Settled populations cultivated the soil as farmers and were familiar with star movements as indicators of the proper times for sowing, harvesting, and for the care and raising of domestic animals. There were individuals among them who possessed special insight into these laws and were able to read "the will of the gods" from the heavens or from the great stone circles that served them as calendars. Temples built into a landscape were so harmoniously adapted to it that we can still sense from the ruins that survive how attuned these structures were to their natural settings. Their proportions were read from the cosmos, and it was a matter of course to orient them in conformity with the movements of the stars.

In all great civilizations, temples were built at predestined points such as springs, or on rock formations of some special kind. But they were also oriented toward the rising point, over a mountain peak, of some bright star. Astronomers have figured out star positions for ancient times and discovered what fixed stars or planets determined such linkings of temples with the cosmos. People of that time felt the cosmos and the divine jointly present in these places. Even such early structures as the megalithic monuments were built in conformity with the cosmic ordering of space and time; simultaneously they provided guidance for the attuning of human activity to the year's course.

We can look at this from a cosmic view as well, for the cosmos was projected both spatially and timewise into the landscape; it was experienced as the scene of the activity of spiritual beings. Thus, what went on in a landscape reflected cosmic activity.

It was because Goethe experienced this on his journey to Italy that he could say as he looked at the art works of antiquity, "These lofty works of art are at the same time loftiest works of nature, produced by human beings in harmony with true and natural laws; here is necessity, here is God."[1] And Rudolf Steiner spoke of the fact that Greek temples simply brought to visibility in the landscape forms preexisting in the etheric life-element, in other words, they traced lines vibrant in etheric space.

There were lines of this kind linking stars and earthly places such as mountains, springs, and rock formations. And when the rhythmic cycles of the planets brought them back again to the same rising point, great festivals were celebrated. Tribes inhabiting the American continent, for example, had a special relationship to the cycles of Venus and to that planet's striking appearances on the horizon; its rhythms and movements were understood to be the "language" of the gods, according to whose dictates earthly matters were arranged. Remnants of memories of a divine landscape—Paradise—and of the Tree of Life and the Water of Life have trickled down to our own time, which has witnessed the final dissolution of such ties. In our liberation and loneliness, however, we have found our way to ourselves, a compensation beyond all reckoning. This has led, of course, to self-sufficient pronouncements such as Lord Bacon's statement that anything having to do with religion—in other words, any connection with a higher realm—is out of place in the investigation of nature. From that time forward, every last discovery and invention was to be assigned its place in the order of things and described solely from the standpoint of weight, number, and volume; it is to be measured, dissected, analyzed,and catalogued exclusively on the basis of our physical senses and technical equipment.

This sort of thing has been going on for a long time, and will continue to do so, though more and more people are raising questions such as, "Should we not reconsider matters? Can the scientific and technological viewpoint really be separated from religious and ethical values without sacrificing human welfare? What is the civilization we have built up around us all about? What is life's meaning? What, indeed, is life itself? Scientifically considered, how did it come into being on earth?"

We receive answers like: "Life is of chemical origin," or "Life is matter, dead matter." One version is typical of all. We cite Georg Kleemann discussing it in the Stuttgarter Zeitung for March 17, 1977:

> The geneticist Carsten Bresch reckons with evolutionary periods and sees intellectual evolution, based on the possession of a brain, as a later successor to biological evolution. The human spirit replaces previously prevailing laws of life with the evolutionary principle of intellectual information and changes biological patterns to support it; the new molecules being produced in chemists' retorts are products of the coming intellectual evolution which will one day no longer permit the survival of what suits nature, but rather of what suits man.
>
> There is a catch, however, and that is man himself. For the pre-condition to this further step in evolution is that he must learn "to think responsibly." When he has achieved this, an intellectual organism will come into being out of the inner necessity of the way things work in our world that will be capable of resolving the contradictions still plaguing us today. With the minor assump-

> tion that man will struggle through to the state of reason, evolution will have its way as a process aiming at the production of every more interrelated and complex conditions of matter.

And Kleemann goes on to say:

> The author thus subscribes to a bold vision of the future and hopes that man will be capable of participating in further earthly evolution, far into the future. And it is here that he parts ways with current biological thinking, which assigns every living being, including man, a beginning and an end; he becomes a heretic who, with a kind of thinking bordering on the religious, regards the entire evolution of the universe as the outcome of its initial physical conditioning (here his colleagues agree). But he goes beyond this to prognosticate an intergalactic evolution encompassing every inhabited planet in the cosmos. On this point his fellow-professionals are certainly not prepared to make any exact statements.[2]

"With the minor assumption that man will struggle through to the state of reason, evolution will have its way. . . ." Reason is manifestly not yet a part of the system; there are as yet not even human beings endowed with reason in it, not to mention angels or any divine element. Does this not mean that all the lights of heaven are extinguished, every connection with anything higher broken off, and the bottom of the valley reached?

What are human beings, then, as viewed in this quotation? The answer can only be that they are groupings of molecules. And what of thinking? About thinking—the common basis for all their striving for insight—the various sciences have as yet failed to reach, or even to try for agreement. What is morality? Again, it is just certain molecular groupings. And this concept has already been put into practice. Approximately twenty thousand operations are being performed annually in Germany, using probes to burn out certain cell areas of the brain in "persons with criminal impulses"—cell areas held "responsible" for moral attitudes.

Further questions related to our theme press for clarification. What is the earth, and how did the earth and the universe come into being? The answer given is that it is a product of chance in the vast emptiness of cosmic space, a collection of elements which it is quite all right to manipulate at will, even if life-patterns are destroyed in the process. A quote along these lines appeared in the *Südkurier* of February 24, 1976:

> An explosion began everything. The universe originated in a tremendous explosion over ten billion years ago. It has been expanding ever since that "big bang" like a balloon being blown up.

It need not surprise us that all sense of an ordered cosmos is missing from hypotheses of this kind. Yet the grandiose harmonies of the *Mysterium Cosmographicum* were discovered by Johannes Kepler as the outcome of observations that provided scientific evidence of the organismic nature of the cosmos, and, as part of it, of the earth as a life-realm. It was a matter of course for Kepler to recognize that here, too, as in every organism, there had to be interaction between the parts, and he had actual experience of the fact that the earth was a living being, with its internal and external aspects in communication, thus with interaction between its parts. This harmonic "orchestra" still so livingly experienced by the Pythagoreans is heard no longer. But this raises the question whether it is not humanity's present task to rediscover that organism of cosmic dimensions and to give it support and care.

Can we expect efforts aimed at a renewal of this kind from a science that wants the ideas of Kepler and his ilk eliminated in favor of such ideas as that of "cell-molecule groupings"?

It is essential to realize that the real causes of the huge problems confronting us in every aspect of social life, including criminality, drug abuse, and terrorism, are not to be sought in brain cell molecules—the source to which the researchers attribute even the products of their own minds. Do not the miseries of the time spring rather from the hopelessly muddled state of the thinking that prevails in the current forming of scientific hypotheses? Must there not be reaction when it is felt that the great truths are to be found elsewhere? That thought-constructions like the above-mentioned no longer have anything to do with the temples of the gods from which the design for the shaping of social life once issued? We certainly realize that humanity's earlier connection with them, which prevailed when human beings lived in a different relatedness to the earth, the cosmos, and thought, has been broken off.

Many people are devoutly hoping to see the problems solved, though they ask from what sources solutions may be expected. Can we hope that these solutions will come from the deliberations of scientific congresses that have given rise to some of the statements quoted here?

On the physical plane, technology has accomplished admirable things. But we note that in many instances it reduces itself to absurdity. There are many imponderables to be considered in dealing with the life-element, and that, in the last analysis, is what all technology should be serving.

The total helplessness as well as the basic orientation that prevails shows up in the following small sample of what today's professionals see as the proper way to make plans for a Black Forest landscape:

> The establishing of further fish ponds in the Hotzenwald is ruled out. Except for areas set apart for settlement, the only interference with the natural landscape that will be permitted will be reservoirs, autobahns, roads needed for intercity traffic, and high-tension power lines.

We know, of course, that Lord Francis Bacon's thesis, "science without religious and ethical aspects" had to bear fruit of this kind. But our time longs for more than science and technology have to offer "as the price of progress." Has not an equally exact science of the spirit that takes the neglected side of existence into account long been a fundamental need? Why is there a turning to yoga, astrology, and mysticism if not to counterbalance contemporary civilization? And are our scientific investigators themselves able to do wholly without "spirit," considering that spirit is encountered everywhere we turn in both the microcosm and the macrocosm?

A large number of phenomena are encountered and investigated daily, phenomena that cannot be explained without taking into account the spiritual element that reigns in nature. As Goethe has Mephisto say in *Faust*, "where concepts are lacking, a word is found to fill the vacuum."

What is the word to which people resort in attempts to avoid speaking of "spirit"? It is the magic word "information." Wherever spirit is encountered and traditional concepts are inadequate, the word "information" serves instead. Matter builds information into itself, in everything living, and in all cellular processes. Groups of molecules are the operative factor here; they pass on information or commands that give purposeful direction to other molecular groupings. Woe to the students who use the word "spirit" or "essential spiritual being" in a dissertation these days! Information is the only help in the dilemma in which those students find themselves. And so we see treatises and dissertations simply swarming with information. Instead, straightforward reference should be made to a science of the spiritual realm researched and elaborated decades ago, for example, in the lifeworks of Rudolf Steiner. In his works, exact and ethically responsible research, which is permeated by a genuine love of truth, can be found.

His is a science founded upon an epistemologically irrefutable view of the reliability of the thinking process. It is set forth in such books as *Truth and Knowledge* and *The Philosophy of Spiritual Activity.*[3] These books are results of introspective observation based on the methods of natural science. It is further explored in *A Theory of Knowledge Implicit in Goethe's World Conception, Evolution from the Standpoint of Reality,* and *Knowledge of the Higher Worlds;* the latter is a guide to extending our capacity for knowledge without retrogressing into yoga or some other form of untimely mysticism.[4] These books provide the basis for establishing a new, scientifically grounded relationship between the earth, the cosmos, and the spiritual world from which cosmic order issues forth.

While still a child, Rudolf Steiner had concrete experiences of that realm beyond the world open to the physical senses. But how to find the language whereby such experiences could be conveyed to the modern consciousness, a consciousness dominated by a science to which only the ponderable is real?

Rudolf Steiner found the answer in Goethean science when he was called to the Goethe Archives at Weimar and given the task of editing Goethe's scientific writings. Goethe had succeeded in finding ways to express spiritually perceptible aspects of experience beyond the discernment of the ordinary senses. Such, for example, was his "idea of the plant," the "primal plant," or "*Urpflanze*," or the concept—only to be spiritually grasped—of the "type" applying to animal species. Then, too, his concept of the spiritual entity "light" as a fact only to be grasped by thinking, and of colors as the "deeds of active and passive light."

This has been the kind of insight and the cognitive methodology we have been at pains to develop in our work here. Earlier lectures, too, have shared this striving. We have tried to learn how to perceive the essential being of water as demonstrated in its "deeds" and in the way it works in nature, creating meaningful order. We see this, for example, in the law-abiding patterns of the "dynamic" meteorological year, of which water, with its special attributes, is the designer. We showed how water can be recognized to be a *being* prone to renunciation; it renounces any inherent form of its own, any color, chemistry, and rhythm—thereby making itself capable of serving as the basis for all organic form, for all the colors of the rainbow, for almost all processes of chemical change, and for the existence of rhythm and movement patterns. Its well-nigh unlimited omnipresence allows it to put its attributes to use in serving life, and its universal transforming ability makes it the matrix of all metamorphoses of form.

We have water's universal nature to thank also for the fact that we can develop our bodily form in the fluid of the maternal organism; that water can serve us as the life element after we are born; and that we are able to think because our brains are afloat on water's buoyancy. It was this universality that caused water to be revered in all the great civilizations of earlier times as the all-embracing mother.

It is facts of this kind that enable us to find our way to a new connection between the earth and the cosmos. The watery element serves as the mediator, and scientific insight represents the first step forward. For it is just water's attributes that point out the path for that insight to follow and the direction to proceed in, namely, dissolving rigid prejudices and fixed opinions, achieving a mobility of thinking suited to understanding the life-element in its essential creativity, and keeping ourselves selflessly receptive on the soul-spiritual plane to new possibilities of shaping and metamorphosis.

In this lie the answers to questions raised by Jesus' conversation that night with Nicodemus: "Unless ye are born of water and the spirit ye cannot enter the kingdom of heaven." That conversation truly sets forth the cognitive preconditions for finding our way to a new connection of earth and the cosmos with the help of the mediating element of water and of the spirit underlying it. It is up to us to make good, with an overdue height-

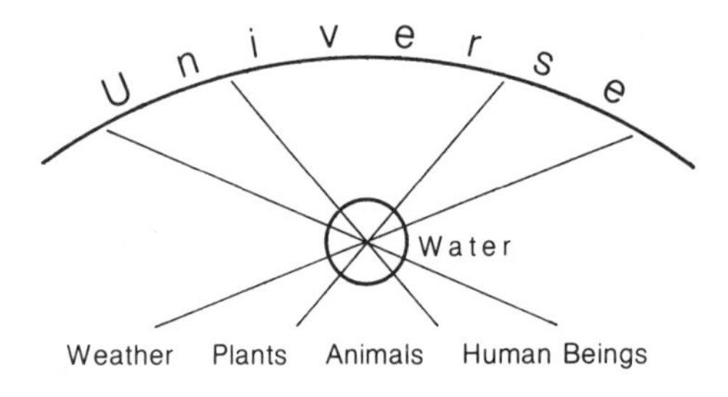

ening of awareness, the past loss of connection with the cosmos.

In an earlier chapter, "What is 'Living Water'?" we presented a diagram intended to demonstrate how all earth's life is born of water and developed by it:[5]
First, we demonstrated the earth's own organism with its rhythms, second, the world of plants, and of animals, and finally that of human beings. All life is the product of the interplay between the earth and the universe with its spirit-created patternings; water is the mediator; the heart organ that lives out its rhythmical being in-between.

But must there not be repercussions when a single organ within the whole of a larger organism is thrown out of balance, as has now happened? And may we not, on the other hand, look for healing effects when living systems are given proper care and maintenance? This has been the theme of all our other presentations here as well.

So we arrive at a new diagram that was developed out of anthroposophical spiritual science, a diagram that has begun to be put into operation—in agriculture,

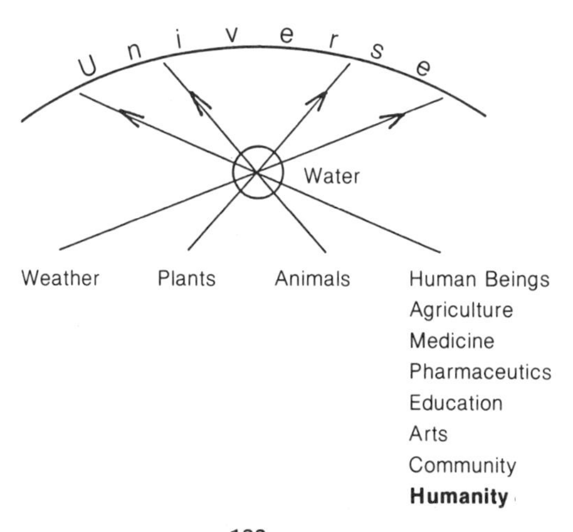

medicine, pharmaceutics, and the study of man, in education, in the arts, and in the shaping of human society so that the earth can be freshly and livingly connected with the cosmos in a reciprocal giving and receiving.

Motion Research: Its Course and Aims over Twenty Years

Theodor Schwenk

The theme of this year's conference brings us again to a concern with weighty problems, problems involving threats to the earth-organism, threats that every one of us encounters daily. Two of these, the themes of energy and of the nutrition of the human race, are themes upon which the attention of the civilized world has recently been focused. Our past conferences have concerned themselves over and over again with these great overriding problems, and the Association for Motion Research has also been participating in the effort to solve them. The Association for Motion Research was founded in a situation and at a point in time when the present state of things could be clearly foreseen.

Let us begin with the problem of nutrition. And let us keep in mind as we do so that this problem cannot be separated from that of an earth-organism in a healthy state of its life-elements: light, air, warmth, soil and water, and, of course, their interaction in a balanced climate on this planet.

Changes are already showing up at great heights in the sheaths or mantles enclosing the earth, changes that are interfering with and leaving their mark upon the

---

Content of a lecture delivered in the summer of 1979 to the members of the Association for Motion Research at the Institute for Flow Sciences at Herrischried, West Germany.

life-processes of this planet and its organisms. Until a short time ago, an intact sheath called the ozone layer was still in existence, but it is increasingly being broken down. This layer, which surrounds the earth at a height of 20–50 kilometers, acts as a protective shield, absorbing the ultraviolet rays of sunlight that pose a danger to everything living.

How may we describe the process taking place there with such powerful though as yet scarcely noticed effect?

What is involved is a chemical breakdown of the protective ozone layer by nitrogenous waste gases and those propellants familiar to all of us from their daily use in almost every kind of spray. They

> . . . guide to their targets in a fine mist and in easily controlled amounts deodorants, paints, insecticides, fluid bandages, waxes, hair-setting lotions, oils, perfumes, and medications. Year after year, all over the globe, many billions of spray cans are being emptied, releasing huge quantities of freon gas. These rise slowly to the highest stratum of the atmosphere, ending up in the so-called ozone layer which forms a protective belt around the earth with a thickness of some 30 kilometers or so.[1]

And there they bring about chemical changes that result in the breakdown of the ozone layer.

> Their special danger lies in the fact that it will be possible to measure their harmful effects on the ozone layer only after the damage has become irreparable. . . . Ozone absorbs the sun's deadly ultraviolet radiation. Thus the ozone layer not only keeps a skin cancer epidemic from

> breaking out on the earth: scientists consider it possible that a weakening of our planet's natural defenses against radiation could lead to a melting of the polar ice-caps and result in a second Flood.[2]

As regards the light aspect, the skin cancer dilemma is already there in latent form. It had, in fact, become manifest by 1981.

More familiar to us from personal experience is the increasing pollution of the earth's atmospheric mantle. Automobiles, aircraft, and industry have already brought about a 10% lessening of sunlight over entire regions—a figure that should, and certainly does, horrify every climatologist.

This atmospheric pollution has a quality all its own. Exhaust gases released by gasoline and diesel motors contain carcinogenic substances that must be recognized to be in causal relationship to the great and rapid increase in cancer of the lungs. So everything that lives and breathes on earth faces the threat of cancer from the air as well.

What the atmosphere absorbs of these substances released by combustion motors, including lead, cadmium and the like, naturally rains on the soil where our food is grown. Here again we are confronted with the cancer menace. In the soil's case, it is not only a matter of the penetration by exhaust fumes, but also by solids such as deposits left by the wearing down of the rubber tires of motor vehicles, which contribute their own particular chemical make-up to the whole. Add to these the inconceivable tonnage of artificial fertilizers with which the earth's soils are being ruined. According to the *Frankfurter Allgemeine Zeitung*'s financial report of January 22, 1979,

> . . . the worldwide consumption of fertilizers in 1977–1978 is reported by Ruhr-Stickstoff, Inc. to have risen to ninety-nine million tons, an increase of 3.1%.[3]

If we consider that the living structure of the earth's topsoil is destroyed when mineral fertilizers are applied, we can form some idea of what to expect from "a tremendous increase of nitrogen fertilizers in the developing countries."

But here we find ourselves back again in the realm of the earth's water-organism, which—because "dogs bite the hindmost"—gets everything unloaded upon it, every last health undermining element present in the air and the soil all over the earth.

I am referring to the cancer situation. If we look around us in our own country, we find the Rhine River daily expediting fifty-six thousand tons of dissolved salts over the border into Holland. Sixty thousand chemical compounds bear witness to the proverbial "price paid for progress." It is a price we are indeed going to have to pay for our own destruction.

Now what does all this signify for water as the life-element?

1. Not just a process of becoming physically heavier, but the loss too of the forces of buoyancy in living water, forces with which every living creature functions;
2. The loss of that sensitive relationship to the cosmos that has so often been the subject of our talks here, and which the findings of our Institute have been able to demonstrate;
3. The loss of the life-element's capacity to mediate the forces raying in from the cosmos to all living

creatures also threatens as a result of the described loss of the forces of buoyancy.

Matters have now reached a point where the element upon which life and health depend must itself find healing if human life on earth is to continue.

How, finally, is the interplay of the elements of light, warmth, air, soil, and water to bring about a state of balance in the earth's organism if every one of these is weighted down with the heaviest burden of tendencies to illness? Need it surprise us that physicians all over the world are saying that cancer has become an illness out of control?

If the damage inflicted upon our environment has played havoc with the elements upon which human nutrition is based, the dangers inherent in the forced building of atomic power plants assault it directly. It has long been known that the incidence of leukemia, or blood cancer, is 30% higher than average in the vicinity of nuclear power plants. And is it not symptomatic that the use of heavy water plays such a significant role in the technology of these facilities? Does it not represent the exact opposite of living water with the latter's forces of lightness and buoyancy? Proponents point out, of course, that the amount of radioactivity released by these plants scarcely exceeds that released by natural geological radiation. But what transparent logic the arguments of this kind employ! Can it be asserted of any serious poison that it is "completely harmless" if a further dose of the same size is added to natural exposure? And what of reactor accidents in which many times the amount of natural radiation is released?

Apart from these considerations, energy so produced

makes use of already stricken rivers to cool the tremendous heat generated by nuclear power plants. Therefore, professionals whose concern is with the life-realm speak of "fever-stricken" rivers that flow downstream from nuclear facilities with their very delicately adjusted living ecological balance destroyed. From this aspect too, rivers must be declared regions of death in their degeneration from the state in which they once served as life-arteries running through landscapes. Ecological adjustments in a river's course are the very elements that go to make up the life-pulse of the earth-organism. Just think, the buildings housing nuclear generators will relatively soon be standing for thousands of years in ruins that are unsafe to enter, and they will have disintegrated under the impact of natural forces long before the radiation still saturating them is exhausted.

Are we not dealing here with images of real monsters of the future that are housed in such facilities night and day, year in and year out, for twenty thousand years or more, to break out in deadly radiation, destroying every remnant of life, or creating the sort of human caricatures familiar to us from the atomic bombing of Japanese cities? These have become decisions of apocalyptic importance, a battling with the dragon for the higher ego and for physical bodies of genuinely human form.

Who is responsible for deciding such issues? No matter how aware the scientists who pioneered the discovery of nuclear forces were of their responsibility—there is talk of "the inescapable necessity of developing nuclear energy," and of "the price of progress"—progress that cannot but bring about the destruction of the life of the earth and of humanity. Does it really spell a retreat into the Middle Ages to postpone further development of nuclear energy until the developers have acquired the moral capacity to deal with it? We must state on the basis of facts that to let these forces loose—for that is what it amounts to when we weigh what has been reported here—means a new Ice Age, not a retreat into the civilization of the Middle Ages. This eventuality was thoroughly discussed in my talk entitled "Keeping the Earth the Place of Life."[4]

How has the situation developed since that talk was given in 1975? Two opposite camps of experts have voiced their common opinion that the life-processes in the earth-organism have reached a state of extreme lability. One camp speaks of the eventuality of an imminent new ice age, the other of a warming up of the earth as a result of air pollution, which reduces the radiation of heat into space. While the latter camp sees the beginning of an ice age as millions of years distant, the former prognosticates the possibility, in the near future, of an upset in the whole system of balances. They cite facts such as:

> All that stands between us and an ice age is 1.6 degrees Celsius (2.88 Fahrenheit). . . . And . . . the boundary separating the peoples of the northern hemisphere from a climatic catastrophe has grown gossamer-thin. An average drop in temperature of 10 or more degrees Celsius (18 degrees Fahrenheit) would conjure up a new ice age. A hardly noticeable drop is enough. If the average temperature goes down a mere 1.6 degrees Celsius, glaciers will begin the long march from the mountains into the valleys and from the poles into regions presently capable of settlement. . . . The earth's ice masses are already expanding. The icecap covering Antarctica is now 10% larger than it was just a few years ago, while that at the North Pole has actually grown by 12%. The first generally

> noticeable effect is the lengthening of winter by almost a month (from 84 to 104 days) in the middle latitudes of the northern hemisphere during the past decade. In other words, frost appeared earlier and lasted longer, and snow began falling earlier and remained unmelted for longer periods. The growing season shortened.[5]

It is not hard to discern what is causing changes of this kind: the life of the earth-organism and of its sheaths is being interfered with by activities of human beings. Groups disconcerted by this recognition counter it, of course, with slogans designed to win acquiescence, such as "the price we must pay for progress," or the threat of "a return to the Middle Ages without nuclear power," or the "inescapable need for nuclear energy."

Let us take note that this is resorting to the kind of talk that characterizes certain spiritual powers in the apocalyptic battle. On the one hand it presents pacification, a dressing up of the facts, a refusal to worry, irony, and minimizing; on the other hand, it points to "the inescapable necessity of nuclear power," with no regard for the consequences for humanity and the earth-organism.

And the sense of haste is such as to raise the momentous question: At a time when so much energy is being expended on sealing the fate of earth and man and where so much darkness is being poured out to put humanity to sleep, must we not look for the advent of great light? Are spiritual decisions not implied in symptoms like these?

We are not unwritten slates in such matters. It has been possible for decades for every seeker to find the anthroposophical spiritual science of Rudolf Steiner. He called attention to the great spiritual decisions pending in this century, decisions that we now see confronting us on every hand in shatteringly palpable reality.

us on every hand in shatteringly palpable reality. Rudolf Steiner pointed to the great event that is beginning in this century, that has already begun, the appearance in the etheric world of the Risen One. He spoke, too, of the concurrent opposing ever-greater concentration of evil, of the caricature of human nature, and of its appearing on the physical plane in the incarnation of Ahriman, the spirit of heaviness, the promulgator of the "necessity of inescapable situations," the spirit of unfreedom as opposed to the spirit of freedom and of human love.

It would be astonishing if calls for a new ethical attitude toward life were not to resound in a situation rife with such stirring spiritual decisions. Examples were cited in "The Spirit in Water and in Man" (pp. 27–38). Let us look again at one of them:

> People have not yet entered into a right relationship with technology. We must learn to think in a different way and change that relationship. Man, whom all our efforts—including those in the field of technology—should serve, must be made the central focus of this new way of thinking. . . .Technology's ingenious accomplishments are the product of the conscious human spirit. But human intellect is not by any means our highest function. Despite the wealth of its ingenuity and its logic, it lacks depth of feeling. It helps us to determine degrees of efficiency and guides us to economic applications, but it has nothing to offer on the score of "good" and "evil"; it does not motivate us to resist the bad. Intellect permits the misuse of technology in carrying on wars, in greed for possessions, and in other evils. Trends like these can only be fought by an inner soul-change on the part of every individual.
>
> The late Albert Betz, an engineer who accomplished such great things in aerodynamics once said, "Seeing

> created for them certainly dims one's pleasure in the accomplishment. One has to ask oneself whether we were right to pursue such research and to do such creating. The answer has, of course, to be in the affirmative. But one thing has been overlooked: we have to take much more care not to let what we create get into the hands of people who lack an ethical maturity commensurate with technological advance."[6]

Let us hear what another such personality—Hermann Oberth, one of the *fathers* of rocketry—has to say. On the recent occasion of his eighty-fifth birthday his accomplishments and personal qualities were noted in the press. Hans Hartl wrote in his biography:

> Just as, in his student days, he had the genius and farsightedness to foresee and plan the space rocket, he now spoke of things that he envisioned coming in the future. An "electric space ship, large-scale mirroring devices, a moon auto, a speed multiplier" were technical concepts that occupied him. He had, in 1928, already theoretically conceived an electric space ship designed to open up the planetary system to man's exploration. In *Man in the Universe* (1954) he commented that for this purpose large space stations erected with the help of orthodox liquid-fuel rockets would be needed to serve as earth-circling bases for electrically driven space ships. But Oberth pointed more than once in conversation to deeper concerns which marked the rocketry professor as a moral philosopher. . . . "I believe that mankind is in grave danger unless moral development keeps pace with technical advances. And a human race not morally intact could endanger the whole cosmos with its technical accomplishments."

These challenges to a new morality are not rare. They could be cited in a considerable number of cases as the comment of responsible atomic researchers in particular.

A decisive point has been reached in the development of western civilization when individuals like these call for the inclusion of moral values in natural science and technology. It signifies a turning away from postulates like that of Lord Francis Bacon who held that science and religion must be separated. Practically speaking, that separation was intensified by the scientific establishment some time ago in the rather different sense that the very opposite of moral values made inroads into scientific research. Terms like "bribery," "statistical data," "expert opinion" and the like suffice to suggest the trend.

A review of the insight of many ethically oriented scientists often raises the question, "What should we have done, then? Since the situation has become what it is, what can we do now? How can a scientist or technologist keep his discoveries and inventions from being separated from him and used for egotistical purposes? Is it still defensible to adhere to that earlier view that held that new findings should be put at the disposal of the world, regardless of what the world does with them for good or evil?"

Ethical considerations obviously require rethinking here if researchers and technologists are not to become involved in situations conflicting with their ethical convictions. There are, however, viewpoints from which answers can be found to such pressing questions of conscience, and these lie in an exact scientific approach to the moral-spiritual world; these viewpoints are present in a science that has been open to everyone for a long time—the anthroposophy of Rudolf Steiner.

This brings us to the same place in our report where, twenty years ago, we asked the question decisive for the founding of the Association for Motion Research. How can our scientific efforts serve the welfare of humanity and the earth-organism? How can this be done without their serving primarily economic rather than therapeutic interests?

Experience gathered in the course of the Association's work has shown that the question was correctly put. Therapeutic impulses dominated the work of the Association from the beginning. As early as 1948–1951, Dr. Alexander Leroi, an Arlesheim physician, proposed the founding of such an association in connection with the development of a cancer remedy. Certain questions arose at that time regarding a practical method for producing the cancer medication, for which detailed knowledge of hydrodynamics was essential, and Dr. Leroi turned to me for advice and practical cooperation. It was possible to solve the problems involved, and that project was concluded.

But the time was not ripe then for a more extensively conceived Association for Motion Research. During that period, I was devoting myself mainly to experimental work at the Weleda in Schwäbisch Gmünd on problems of the effectiveness of potencies in medicines. This had to do with the discussion that came up again and again of the perennial question, What was really involved in the potentizing process?

There are, as you know, three aspects to this procedure: (a) that of ponderable matter, (b) that of the neutral potentizing medium, and (c) that of the rhythmical shaking that combines them. The experiments referred to led to the publication of my paper "The Basis of Potentization Research," presenting a finding that may be of interest to this gathering.[7] The cosmos with its ever-changing configurations gains entrance to fluids in motion, and in particular to water. This holds true in the case of grinding processes as well.

This finding fully confirmed a statement made by Rudolf Steiner many years ago to the effect that attention should be paid to "the proper moment" in the manufacture of pharmaceutical preparations. At that time, the fact that the cosmos has entrée to fluids in motion aroused strong opposition in some quarters, but in others it met with understanding and insight into its implications for pharmaceutical practice. Today, this fact is taken for granted due to the notable confirmation proffered by the "drop-picture method."

Research led to the discovery that the cosmos affects not only the entire realm of fluids in motion on this planet, but that it also does so in a great variety of ways. This provided a bridge for the linking of the cosmos and the watery element. Frau Lily Kolisko's pioneering work on solutions of metal salts in water, using the *capillary dynamolosis* technique, had long since established such relationships, and other related discoveries were achieved by anthroposophical mathematicians working in the realm of the new projective geometry. We think here in particular of the work of George Adams, who, with his publications, provided the conceptual mathematical basis for linking cosmic activity with form in nature's terrestrial kingdoms. Projective geometry is as though predestined for the investigation of these relationships between earth and cosmos, for its constructions also include infinitely distant points and planes. George Adams, ably supported by his colleague, Olive Whicher, led the way to insight into the world of plant forms in connection with the suction-created buoyancy of the "etheric sun-space" to which plants, countering gravity in their growth, respond.

Practical aspects illuminating the relationship of projective geometry to hydrodynamics came to light in further cooperation with the mathematicians, one of whom was Dr. Georg Unger. Bridges were built from these two differing directions—the experimental and the mathematical—to an understanding of the way the cosmos works in the earthly realm. What could have been more natural than to connect the two approaches—telling oneself that it must be possible to induce movements in water that, in their development, correspond to projective-geometric forms, taking on an organic lawfulness that, according to our previous demonstrations, was of cosmic origin?

The negative influence civilization was having on the watery element, with the implied challenge to find healing measures, was another reason for urgency with respect to these insights and questions. Briefly stated, the problem was, "How can dying bodies of water, and especially the 'dead' drinking water supplies of big cities, be restored to a living, organic lawfulness and spring-water freshness?"

After consultations with participating scientists, particularly with Dr. Alexander Leroi, Dr. Hanns Voith, and Mr. Arthur von Zabern of the Weleda, Inc., these scientific circumstances as well as personal ones, and the external water situation led in 1959, just twenty years ago, to the founding of the Association for Motion Research. The Stutzhof at Herrischried was acquired after suitable preparations and, on June 21, 1961, the opening of the Institut für Strömungswissenschaften was celebrated.

In the ensuing period, the mathematicians worked on the development of the projective-geometric models. I was mainly occupied with experimental work in hydrodynamics, and John Wilkes, who had recently joined the staff, was in charge of the technical production of the models.

An immediate start was made on the first experiments, in which the organic laws of which we have been speaking were induced in water. It was possible, using the drop-picture method, which had already proven itself, to test the products as we went along, tracing the changes in water quality that resulted from the introduction of these movements.

In this connection, we should add that the drop-picture method was especially suited to determining the subtlest qualitative differences in a great variety of fluids, including the effects produced in them by cosmic constellations. The book *Bewegungsformen des Wassers* (as yet untranslated), which was published at that time, contains examples of the pictures. The key finding here was that the *quality* involved actually found expression in the *type of movement* of the fluid medium under study.

This connection between quality and type of movement brought up the question, "Can the quality of drinking water be influenced and improved by the reverse process, that is, by introducing correspondingly selected movement?" But as I have reported, this question occupied us from the moment we began relating hydrodynamics and projective geometry. This fundamental aspect, or the idea behind it, now found full confirmation.

After two years of intensive cooperation, we were

deeply saddened by the death on March 30, 1963 of our much-loved friend George Adams.

The situation was now the following: a series of models based on projective geometry had been designed, produced, and tested with the drop-picture method. Publications included:

- *Basis of Potentization Research* by Theodor Schwenk
- *The Plant Between Sun and Earth* by G. Adams and O. Whicher
- *Sensitive Chaos* by Theodor Schwenk
- *Bewegungsformen des Wassers* by Theodor Schwenk (this book is not yet translated)[8]

As time went on, John Wilkes developed—first at Herrischried and later in England—models for the bowl-like water receptacles that induce lemniscatal motion, which is a self-steering motion.

The drop-picture method continued to show itself capable of reflecting subtlest qualitative differences, including cosmic constellations; it came into use in a number of laboratories.

A certain problem was encountered in relating projective-geometric forms to hydrodynamics. These mathematically developed forms represented what might be called an "ideal or conceptual element" with which water's terrestrial attributes, its weight and viscosity, and its centrifugal pull do not accord. It is just water's distinctive characteristic in its role as mediator between earth and cosmos that it participates in both realms. The ideal forms would have had to take the terrestrial realities into account. Those at home in such matters will be familiar with the relationship of this problem to that of hydrodynamics, where one finds a difference between the pure potential-flow uninfluenced by viscosity and the real potential-flow, which *is* under its influence. The work on the mathematical models did not, therefore, continue in the same direction as before, but from that point onward our research pursued another course instead, one that emphasized purely experimental experience and took actual physico-hydrodynamic realities fully into account. This by no means simple path proved fruitful. We will return to a discussion of it later.

The goal, of course, invariably remained the reenlivening of big city drinking water by means of organic movement, through the incorporation of organic laws, in other words, of etheric forces.

Already at that time, the many studies made with the drop-picture method of big city drinking waters showed the extent to which the problem of water quality existed despite the great efforts made by city water departments. Challenges to responsibility and reverence for water went either unheard or unheeded. Many samples of big city drinking water failed to demonstrate that characteristic balance between water's earthly and cosmic attributes found in wholesome spring water, which has often been the subject of discussion here. What does *getting out of balance* really mean?

Three characteristic types or classes of drop-pictures were constantly encountered in the many studies of drinking waters, of mineral waters from various spas, of medicinal herbs (sometimes with data on their effectiveness as remedies for the patient involved), and, in cooperation with their doctors, of blood samples of patients. These could be correlated with the three organ systems of the human body: (a) the metabolic system, (b) the rhythmic system in its breathing and circulatory aspects, and (c) the nerve-sense system. Experience of this kind provided a reliable basis for judging the quality of drinking waters and determining any one-sidedness they possessed while exploring the interaction of the three basic functions with regard to balance. The book *Sensitive Chaos* demonstrates the fact that good spring water necessarily possesses these three functions: (1) it is suitable for the metabolic area of living creatures; (2) it has a pronounced sensitivity to exceedingly fine qualities in nature; and, (3) it has the function of the rhythmic system. This system is the one most familiar to all of us, for it is witnessed in waves and vortices, in the sand ripples seen on ocean beaches, in processes of a cyclic nature, and in the pulsation of body fluids in living creatures. So it need not surprise us that aspects of the aforementioned balance, or lack of balance, are able to manifest themselves in the way water moves. Viewed from just this angle, the facts justify describing water as *the blood of the earth.*

In applying this diagnostic method to water itself, a hard and fast scheme is not the right approach. What is required here is the kind of diagnosing practiced by physicians, keeping in mind the possible great variety in the matter of symptoms and the connection between them, and combining a great deal of experience with the openness and mobility that the situation calls for. This means that the person in charge of a facility for the reenlivening of drinking water has to become a diagnostician in dealing with the blood of the earth. Such water must obviously first have passed the tests imposed by hygienists. The requirement made of hematologists in their field must also apply in the case of these doctors for water, namely, an exact knowledge of water in its similarity to a complete functional human being, possessing every possibility of deviating from the archetype. The water diagnostician must, in other words, learn to read and to think in the realm of those functions encountered in the drop-picture movement patterns.

The question as to what constitutes a rational therapy naturally arises at this point. As we answer, let us be clear as to the ground we have come to stand on in undertaking specific diagnoses.

1. We can see water as a complete functional human being.
2. The three organ systems are continually present in flowing movement, appearing and disappearing, and can be rendered visible by certain techniques, one of which is the drop-picture method.
3. These organ systems come into manifestation and disappear again as concrete organ forms in motion in an actual flowing anatomy.
4. Every organ in the human organism is originally born of water: the forms of the joints, of the limbs with their convoluted bones, of heart and blood vessels, of ears, brain and sense-organs—the forms, in short, of all three functional systems. They all come into being out of movement and out of streaming currents of the most varied kinds. Preliminary form-stages often disappear, only to reappear in fresh modification.
5. In terms of function, we can see in water a reflection, so to speak, of the human being.

The question of enlivening drinking water in general is related, of course, to that of the mathematical models previously referred to. Our intention was to use them for the purpose of introducing organic laws in general into water.

Since it had meanwhile become obvious that drinking waters have to be restored to health, that they are in fact in need of specific cures, it became necessary to make exact differentiations in the above-mentioned models to match the diagnostic findings in each separate case. To achieve this technically in every case exceeded the bounds of the possible, since there was the added problem of the models' rigidity. We therefore decided to produce models out of water itself when specific organic lawfulnesses based on projective-geometric planes were to be imparted to water. To do this currents were produced made to interact right in the water itself. This conformed to a procedure that we can watch nature carrying out everywhere as it creates organic formations. These currents are attuned to the etheric element and can combine with corresponding currents produced in water; in other words, they are receptive to formative forces raying in from the cosmos.

Once the idea is firmly grasped, it seems therefore to be just a question of experience and also of what technical measures to apply in order to produce these concrete patterns of movement in water by means of water.

We are often asked to satisfy a wish to *understand the rejuvenating setup*. I believe that what I have just described provides some understanding of it in the sense of having also opened a perspective on the working in of cosmic entities. They are that *symphony* experienced up to the time of Johannes Kepler as the harmony of the spheres, and as the cosmic Word "out of which everything was made that has been made."[9] But it is essential nowadays to keep in mind that the cosmic Word passed through the Mystery of Golgotha and since then permeates the earth.

This contributes a vital, indeed, *the* vital, element to an understanding of the water-reenlivening process, for such an understanding is based upon realizing what the

Mystery of Golgotha achieved for human beings and for the entire organism of the earth. So, to answer the question so often put to us, we will quote what the spiritual investigator Rudolf Steiner said about it. Steiner speaks at length on the subject in the lecture entitled *The Etherisation of the Blood: The Entry of the Etheric Christ into the Evolution of the Earth.* This lecture was given in Basel on October 1, 1911.[10] Steiner said that when a person in the waking state is clairvoyantly observed today, light-rays can be seen constantly streaming out of the heart region toward the head and quivering around the epiphysis, the pineal gland in the brain. This process springs from the fact that the blood continually dissolves itself into etheric substance, which streams upward and encloses the epiphysis. It is the intellectual element that, in our waking state, streams as described from below upward. There are other streamings moving counter to this, from above downward, in a sleeping person. These latter come specifically from the cosmos into our inner being and represent a moral rather than an intellectual element. As Rudolf Steiner described it, it is particularly the "moral quality" of an individual observed by the clairvoyant that is streaming into him in sleep.

Steiner goes on to describe how a process took place in the macrocosm similar to that of the transformation of material blood into etheric substance. It was the event of the Mystery of Golgotha, when blood from the wounds of the Redeemer penetrated the earth. In the earth's ensuing evolution, the same thing happened with this blood that happens also in the human heart. It underwent an etherizing process, so that ever since then the etherized blood of Christ Jesus is present in the ether body of the earth.

A connecting of these two streams in the etheric bodies of the earth and of human beings can take place when individuals make the effort rightly to understand what happened on Golgotha in the way made possible to us by anthroposophical spiritual science. It lies in everyone's free will whether he will connect himself with this stream of the morally good and thus make it possible for both streams of the etherized blood to be united.

If these communications of the spiritual investigator are received with understanding and thoroughly pondered, they provide insight into the spiritual and—surprising as it may sound—the concrete bases for the conception of a reenlivening process for water, the blood of the earth. This is the aspect from which—and from which alone—our idea for enlivening water can be understood. It is just this kind of understanding that has gradually to be acquired, and it is especially essential for those in charge of water-enlivening plants to develop an understanding of what is taking place there invisibly and can take place ever more livingly as this understanding grows. It is this same understanding that must come to be included in our modern scientific knowledge of the working of the cosmos in flowing water and that indicates the great responsibility which physicists and scientists of the future must develop.

If this were accomplished, the warnings voiced on ethical grounds that technology must not be separated from moral values and that technologists must remain morally connected with their discoveries in order that these serve human welfare would be satisfactorily met. With recourse to Rudolf Steiner's spiritual science, ways can be found to practical realization of these ethical challenges and their effectiveness can be demonstrated.

In *Microcosm and Macrocosm,* originally given as lectures in Dornach in May 1920, Steiner speaks of concrete aspects of this kind.

> Christianity will not have been properly grasped until it has permeated the earth right down into the physical sciences. It has not been understood until we grasp, right down to the physical level, how Christian substantiality works throughout the entire life of the universe . . . .
>
> The spiritual world view must merge with the natural outlook on the world. But this occurs *within man,* for they merge through the doing of a free deed.[11]

The question presently weighing on so many individuals who are aware of their responsibility is the same that these warning voices might also have expressed: "What should we have done to prevent what has now happened all over the earth?" An attempt to indicate the answer to this was made above: Seek for a new, timely understanding of the Mystery of Golgotha such as has been available since this century began. Then new ways will be found—ways reaching right down into the "realm of physics."

However, we see that to understand a water-etherizing plant is, at bottom, to understand the event on Golgotha of which Rudolf Steiner spoke.

The fact that a perspective like this, opening onto a Christianized, ethically oriented physics and technology, is offered by anthroposophy could also be the answer to a further inquiry as well, one that we often meet with. It is: "We would really like at long last to *see* the machine for the enlivening of water."

To this we reply first with the question, "Can an enlivening process ever be just a 'machine,' a machine that might conceivably be misused in what our time thinks of as profit making enterprises?" True, the enlivening process uses materials, earthly materials that must be worked on with vise, lathe, drills, and polishing machines. But the *processes* that take place in our conception of an enlivening plant correspond to those of *organisms*. Just as little as we can compare the heart organ where the etherisation of human blood takes place with a pump, so we cannot regard the "heart-portion" of a water reenlivening plant that is to treat the blood of the earth as a machine. The technology involved in such a procedure is not the primary consideration. What *is* primary are those invisible processes without which transsubstantiation into the living element could not take place.

In addition to the above requirements for this kind of organotechnology, invisible helpers are needed, helpers who in earlier times were encountered in nature, namely, co-workers in the creation of all living form in earth, in water, in the atmosphere, and in warmth. The beings referred to are the elementals, who, in the words of Saint Paul, are the "creation" that longs for man and is waiting for him.[12] These beings are to be thought of as included in the dedication spoken by Dr. Hanns Voith, the first chairman of the Association for Motion Research, at the opening of the Institute for Flow Sciences in 1961.

> May friendship and trust
> Reign in the individuals
> Who work in this building.
> May the elementals of the mountain springs
> Incline themselves in friendship to this work,
> And then to the inquirer,
> Reverently approaching nature,
> Will be revealed what now is hidden.

Water as the Element of Life

Theodor Schwenk

Looking back into the distant past as far as our awareness reaches, we find that water was the object of human veneration, a veneration amounting to religious worship. Every great culture felt water to be connected with the loftiest gods. It was considered holy, an element not to be tampered with in its purity. There was, though, even in the early days of Greece, a revered figure who bewailed the fact that it had become necessary to establish laws for the protection of this substance previously held in the highest regard as a matter of course. That was the first step in a lowering of standards that has continued, with only occasional interruptions, right down to our own time. Plato (472–347 B.C.) saw it coming, and he warned that the forests of Greece were being decimated, and that this would bring about drastic changes in the water situation in his homeland. What would the civilization of Greece have been without its lakes, its sacred groves, and its springs and rivers? Its decline definitely began with the deforestation of its hills and mountains.

The lesson we can learn from this example has been repeated more or less distinctly in the case of every great civilization. A lessening of reverence for the divine

---

Translation of an article in *Soziale Hygiene: Merkblätter für Gesundheitspflege im persönlichen und sozialen Leben*, 1982.

is reflected in a diminishing of the life-element for the human race. Felled forests and neglected water sources seal a civilization's fate.

What experience of water can big city children have today? They turn on a faucet and see water pouring out of it; it disappears again down the drain of the sink. Or they can cavort in the bathtub, often enveloped in a reeking cloud of chlorine. What a joy it was for city children, just a few decades ago, when a cloudburst brought such floods of rain that the sewers could not take care of it! It would be dangerous today to come into too close contact with this kind of water.

What is the situation with brooks and rivers outside the towns? Signs forbidding bathing sprout like mushrooms. When a townsman goes to the beach and is sitting reading his newspaper, he is sure to come upon big headlines reporting catastrophes such as oil spills, fish dying, and inadequate water supplies.

A few years ago, when some prominent people were asked to describe what they considered the ideal vacation, it turned out that the great majority yearned to be near water, at the ocean, sailing, fishing a clear stream, at a waterfall in the mountains, or in a forested valley where springs bubbled and brooks ran. Such retreats enable people to experience the marvel of renewed life and energy, even if it is not susceptible to scientific proof. People are drawn to these situations even though the professions to which their ordinary lives predestine them result in technological damage to living interrelationships in nature. It is not just the change of air that proves refreshing, but a change of water too. People feel quickened when they drink from a cool, clear spring, and they gather new strength from seeing clear water coursing merrily along. What an opposite, sickening effect is felt at the sight of dead, turbid water in an oily, sluggish flow!

Water is predestined by its possession of the regenerative powers described to be the life-element par excellence, a nutritive element for plants, animals, and human beings. We are, of course, familiar with the fact that growth, nutrition, and the circulation of the blood are based upon it, and that its stimulating motion also gives nourishment to those senses whereby the flow of life is quickened in us. What other explanation could there be for the above-described experience of rejuvenation?

Now that we have taken a look at these imponderables, let us call to mind some of the ways in which the life-element expresses itself. All living things have in common growth, reproduction, nutrition, excretion, metabolism, changes in form, the ability to regulate warmth, chemical processes, osmotic pressure, processes of becoming and decline, and, last but not least, the rhythmical processes that pervade all life-functions. Readers must consider as they peruse this how all these functions of life depend upon water. Then, they will be able to confirm that none of them are conceivable without the presence of water as a mediator. Every such process can take place only with the help of water; it is the essential, interrelating primal substance in all life-processes, everywhere present, and serving flowing life. It renounces fixed form, remaining fluid, and thus comes to symbolize life itself. In sacrificing any shape of its own, it becomes the agent of all other shaping and of all life-processes. Thus, it is a participator in all life's secrets, and a study of life is always necessarily a water-study. Life is an ever-flowing, ever-changing element like the watery element that enables life to come to manifestation in the visible world. Water is certainly not *less* than a living being, since it is the mediator of life-processes in all living things, ranging from the least to the most complex organism. It contains them all potentially, encompasses and suffuses them, and is as it were the primal material substance of existence.

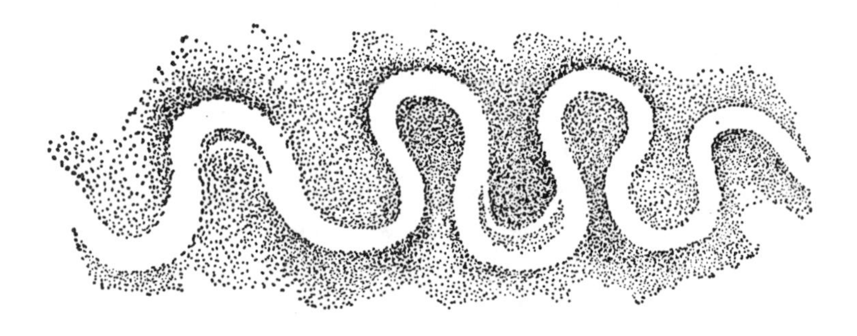

Figure 1. Naturally flowing water always seeks out a meandering course.

Let us ponder the fact that the life-processes so widely disseminated throughout nature are to be found concentrated in the human organism, appearing there in threefold patterning as the limb-metabolic system, the breathing and circulatory organism, and the nerve-sense pole.

The specific life-processes of these three bodily regions also depend on the watery element, a fact from which we can draw the conclusion that water adapts itself to them in a threefold manner: supporting metabolic change, mediating rhythms, and providing the foundation upon which the organs of perception base their functions.[1] No material change could actually ever occur in nature without water. A study of the sense organs teaches us that water must be present here, too, if the perceptive functions are to take place in a healthy manner. And what of the rhythmic element? It is no less dependent upon water.

But the life-processes in many living creatures are also recognized increasingly as tied to rhythms in the cosmos. It is said that living creatures have an inner clock, functioning with inconceivable precision in some water creatures, for example, and keeping pace with cosmic processes. It is just in the case of creatures of this element that such correspondences are most frequently encountered.

Who acts as prompter in these cases? Could the watery element itself be the mediator for cosmic rhythms? Do the tides not rise and fall in the moon's rhythm? Was there not always an awareness of such relationships in every great civilization, as when people spoke of divinities inhabiting bodies of water, or of an angel stirring the water of a pond into rhythmic motion from which healing resulted?

Descriptions of water would tell only half the story if this cosmic rhythmic aspect—an aspect also found in our whole rhythmic organization—were to be left out.

Awareness of the relationship between the rhythmic element in us and in the cosmos has largely been lost. But awareness of water's ability to take up cosmic rhythms and mediate them to all living organisms has also disappeared. This means that people in our time lack essential insight into water.

When a technician omits some essential factor from his calculations, it is not surprising that his structure collapses. In our dealings with water, some of its essential attributes have not been taken into account, with the result that nowadays water management and balance in nature tend increasingly to break down. If this trend is to be halted, there must be a general realization that water is not just a substance, but a participant in cosmic processes. Teaching on the subject must be included in school curricula if a collapse of nature's balance is to be avoided. Insight into water's total being has always been the foundation of true human culture.

We will return to this subject. But we ought first to look without preconceptions at the water situation as it has recently developed.

The Present Water Situation and Its Consequences

Let us look first at the quantitative aspects of the problem.

Last century, in Goethe's day about thirty liters of water were used per day per person. We have to picture this amount being fetched in pails or pitchers from a spring or well. It was certainly a very personal experience to have to go on foot, perhaps twice daily, with a bucket in each hand. That brought the carriers into direct relationship with the watery element, and taught them how to handle the situation to prevent spillage.

Today's requirement in Germany is for 150–200 liters per day per person; therefore, we would each have to go to the well eight times daily with a 10-liter bucket in each hand. In the United States, the daily consumption of 450 liters would mean making twenty-two trips to the well and back. But we can get all the water we need out of our faucets without giving the matter a thought or making any effort. We have grown richer from the quantitative standpoint, but poorer qualitatively and humanly. We need not describe here what a person's experience of water used to be as it was observed running out of the well pipe into the collecting basin, its rippling and gushing and murmuring inviting the hearer to a few words of friendly exchange. We should keep in mind, in reviewing today's water needs, that industrial development does not stand still; much more water will be needed in the future. It is expected that worldwide industrial production will more than double in the coming decades. However, nature supplies a constant average, proportionate to the earth's capability. If we reckon realistically, dividing among earth's population the amount of water used by industry, another set of figures turns up. Under those circumstances we would have to reckon individual use at approximately 1500 liters daily. A simple calculation will show that an annual increase of 3% would mean a doubled use of water by the end of this century, while what nature supplies will remain the same. These figures bring home the fact that water will be in increasingly short supply, and this has to signify that problems of water quality, already acute, will become extremely serious.

Now let us look further at the qualitative aspects of this problem. Since surface waters must increasingly be resorted to, we should start with them. What is the situation with regard to our lakes and rivers? The Rhine river is a drastic but by no means exceptional example. It carries approximately 50,000 tons of sodium chloride across the Dutch border daily. This is an amount that corresponds to fifty salt-laden trains of fifty freight-cars apiece. Add to this a quantity of sludge—about 7,000 tons a day. At Mainz it measures in the neighborhood of 20 grams per liter. It should not surprise us that divers at occasional shipwrecks must make short work of it because underwater visibility extends no further than 20 centimeters (on the lower Neckar). At Cologne, on the Rhine, it extends far less than a meter, about one-half of a meter on the average. One can stand on the banks of the Rhine at Bonn and reflect that one is looking into water laden with sewage generated by 30 million people as well as with salts, oils, pesticides, acids, and the like. In view of facts such as these,hydrologists rightly speak of a chemical inferno that affronts not only the onlooker but anybody who has retained a sense of smell.

Lakes, those vital organs in nature's management of life, organs once known as "the eyes of the gods," have suffered extreme damage in their life-structure, for they are still being thoughtlessly polluted by effluents. A daily total of 55 million cubic meters is currently being poured into West German lakes and rivers, 37 million cubic meters of which are either insufficiently filtered or not filtered at all. They pour in day after day, hour after hour. These 37 million cubic meters correspond to an hourly burden of 2 million cubic meters.

This is a process by no means limited to the situation here in Germany, in our own country. The Volga, for example, carries a load of 12 million cubic meters of wastes into the Caspian Sea every day. The rivers of Siberia, famous for their profusion of fish, are presently involved in a death-process as the result of being overloaded with waste-water, which flows into these streams to an extent that makes them incapable of self-cleansing.

It is essential that we familiarize ourselves with facts such as these, for the existence of the coming generation is already seriously threatened.

The question of the oceans as reservoirs from which drinking water supplies might be derived takes on a special aspect in view of the situation just described. Although such a scheme for obtaining food from the sea is already being talked of as the inevitable solution for the future, floods of toxin-laden sludge are pouring into the oceans twenty-four hours a day. Layers of this substance have built up several meters high in some spots along the coast of the North Sea. The flow of muck from the Elbe in the area of the North Sea island Helgoland is plainly visible. Almost inconceivable amounts of wastes from shoreline sources are dumped by ship into the North Sea; 85–90 million tons were the official estimate for 1978. Whole industries discharge their "special wastes" directly into the ocean. A huge company in the United States has been expediting 2 million tons of a weak acid solution into the sea, and we could cite similar instances taking place in Europe.

This development threatens to upset the biological balance of inland and offshore waters. According to the 1980 report of the German government's Council of Experts on Environmental Problems, the North Sea "as a whole is ecologically endangered." Reactor technology requires inconceivable amounts of cooling water for nuclear power plants, and water is taken from major rivers for this purpose. A facility of this kind now being planned for the middle Rhine region, for example, would require a flow-through of cooling water equivalent in size to the Neckar river. This water is heated, and every degree of increase in its temperature means an interference with the sensitive biological balance of a river and with its self-cleansing capacity. Experts have reckoned that at the current rate of development, rivers in the United States will reach the boiling point sometime in the 1980s. Veins through which life once flowed would thus be turned into death-dealing streams.

Ground water must be considered when it comes to planning sufficient water supplies for the future (we are speaking here chiefly of drinking water). Forecasts for the period up to the year 2000 predict that all available ground water will have been used up, and almost 50% of the available surface water. Added to this is a fact that very few individuals recognize. At the current rate of use, the ground-water level is sinking alarmingly. This is the result not only of the heavy domestic and industrial demand, but also of interference with nature that is only now, some decades later, beginning to have visible effects. An example is the straightening of the Rhine, that much-acclaimed engineering feat of the past century. It now appears that nature has been hurt, and not helped there. The straightening in this case, as in countless others of the earth's rivers, has meant that although flood waters could be carried off more quickly,the ground-water level was lowered by the river's sucking action. An example is the correction of the upper Rhine valley, where lowering of as much as 30 meters has occurred in places. This has brought dearth and desolation in its train.

Even more serious concerns than the dwindling of the precious treasure of ground-water reserves face us today in our irresponsible pollution of the remaining ground water as a result of dumping, the penetration of the soil by chemical wastes from domestic, industrial, and business sources, by oil, by the dissemination of chemical fertilizers, especially nitrates, by highly toxic pesticides and insecticides, by estrogens and other hormones, by antibiotics contained in animal feeds, by heavy-metal compounds that are being deposited by contaminated air, dissolved by acid rain, and then washed down into the ground water, and by sewage-laden brooks and rivers.

The prospect so tellingly sketched by Georg Berkenhoff in 1968 in the *Zeitung für Kommunale Wirtschaft* has become reality: "If we do not succeed in protecting our precious water supplies, the industrial revolution will have devoured its children." But business is still exploitation-minded, with the excuse of maintaining prosperity.

Tasks for the Future

Problems of these kinds, which have now brought into question the very survival of human civilization in its entirety, cannot be solved by a few superficial efforts to repair the damage. Healing measures can be introduced and healthful conditions restored only if communal life and action reorient themselves in the sense of Rudolf Steiner's threefold social organism on the one hand, and if, on the other, the earth is conceived as an organism from the scientific and spiritual-scientific viewpoints developed by him—a reorientation based upon spiritual insight into human nature.

But efforts and accomplishments to date should also be reported, as the start of healthy developments in the field of hydrology.

At large conferences and fairs for hydrologists, such as "Wasser Berlin," facts such as the following are already being recognized and reckoned with by a considerable number of scientists:

1. The water organism of the earth is a single entity; river systems must be recognized as living organisms and treated as such. Nonpolitical management of spatial matters over the whole earth has to be achieved. National borders do not have any bearing on water and its management. Everyone involved can only benefit from spatial and hydrological regulatory revision. This would mean, for example, coming to exact

agreements on managing the water of single regions as well as of the whole.

2. Regional management, as here envisioned, calls for deliberate work and planning that would include the time element also. One misconceived example will serve for many: "In view of the multiplicity of unsolved problems, the desire to make use of the sea as a general disposal area continues to grow. An unchecked carrying-out of suggestions of this kind would spell disaster unless international regulations were to be set up. . . .Unthinking application of seemingly harmless current measures means issuing a check that can never be cashed," said Professor Balke at the *"Wasser Berlin"* Congress of 1968. Examples of this kind show how essential large-scale long-term planning is; at the very moment when the sea is being talked of as the solution to the problem of feeding an ever-growing population, plans are afoot to use it as a dumping ground!

Some of the agreements Professor Balke called for are already in existence, but they favor business interests in merely legalizing instead of ending the present situation and procedures.

3. Spatial as well as temporal aspects therefore make worldwide cooperation necessary: "Agreements between nations on the management of water, mutual consideration, and common concern, since water is everybody's fate." This assumes goodwill and a readiness to make common cause. We see here that what is needed is an organization of the whole human race which,

> putting politics aside, takes on comprehensive caring for humanity's *life.*

Are these not lessons that water itself teaches us? Everyone can contribute to solving the water problem. Beginnings in new directions are always undertaken by individuals. If we could pay attention to what the efforts of the hydrologists are telling us, we would find examples of genuine solutions for the future. Here again, one example will serve:

> The hydraulic engineer must have close ties with nature if he is really and insightfully to suit his works to the environment. This is not just an ideal requirement; it has decisive and practical significance. Any aspect of hydrological engineering that has to be pushed through contrary to nature costs significantly more to build and maintain. The strong appeal of this profession lies not just in the designing, the solving of construction problems and the carrying out of projects. This is a profession that calls, in addition to full personal application, for prolific ideas, for an ability to feel one's way into the processes going on in nature, and for a grasp of interrelationships both in nature and in economic aspects.
>
> (Maisch, Assistant Department Head)

And what is required of him applies to everyone else as well. We should all acquire the above-described capacities, for every one of us is more or less implicated in technology's interference with living nature. But we all have opportunities, too, to study water's being in natural settings, even though it is difficult today to find clear, wholesome water without searching for it. We should go out with children and let them have an experience of water in a brook as it runs over rocks, leaps up in wavelets, forms vortices, scatters drops, gleams silver, ripples, purls, and murmurs under the roots of trees along the bank. One should show the children how water creates forms in the sand along sea beaches, in wave-worn cliffs, in ice, and after rains in springtime; how pollen floats in every puddle and forms patterns there; how water turns to mist in falls; and how the sun shining on it makes a rainbow. The more opportunities to observe we are given, the keener grows our sensitivity.

The reverence for water that people once felt as a matter of course and that was reflected in the way they treated it must be reacquired today and actively applied.

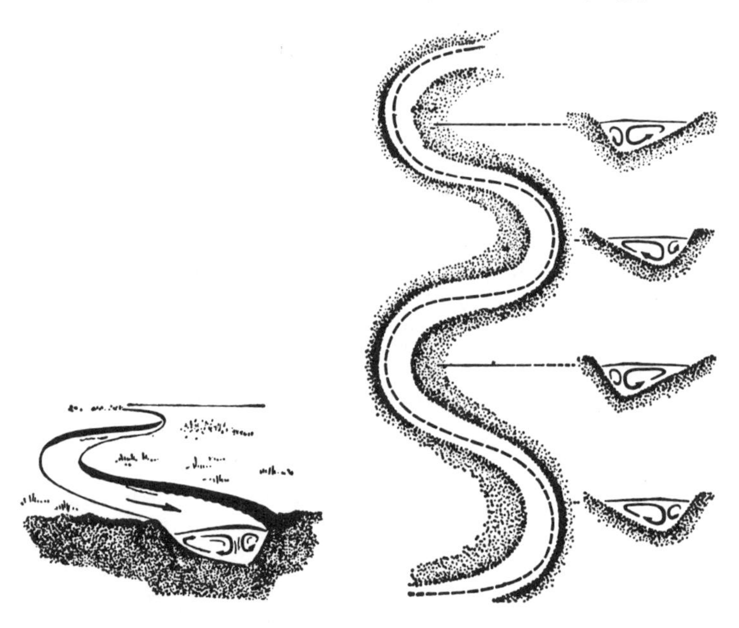

Figure 2. In a riverbed, the secondary rotating currents are embedded in the main current.

Figure 3. The secondary rotating currents vary in size. The larger one adjacent to the flat bank becomes the smaller one, which gnaws at the steep bank.

The demands currently issuing from water specialists culminate in a call for a reborn *water consciousness*. If we do not develop it, the right care and handling of the life-element water will also suffer neglect.

Since the water consciousness we have in mind is only just starting to develop, water must be protected, conserved, and allocated by appropriate laws. The very first laws recorded in history were made for the purpose of protecting water; they were water regulations. They became necessary to the continuance of social cooperation, but only at a period when reverence for water had begun to wane, or where its worship was forbidden as a pagan practice. The present water emergency could be the means whereby humankind is brought again to revere this wisdom-saturated element. Reverence and a sense of responsibility are clearly no longer gifts we

Figure 4. Vortex funnel

are born with; they must be built on understanding and deliberately cultivated—in our case based on insight into water's nonmaterial attributes as well. That is the factor left out of the reckoning by science, as was stated at the beginning of this article. It is a factor that needs to be discussed again here, since it will inevitably play a decisive part in the production of drinking water.

Is Water Just a Substance?

Is water that has undergone mechanical and chemical treatment and had most of the bacteria, chemicals, and adulterants it contained removed to the point that it meets prevailing hygienic standards, and is adjudged potable, really comparable to fresh spring water? On the basis of the experience and insight we have gained at the Institut für Strömungswissenschaften at Herrischried, West Germany, the answer must be a resounding No! Any such water must be regarded as *dead.* A new method developed here—the drop-picture method—that studies the way water flows and the flow-forms created by its movement, and uses these to ascertain the degree of its aliveness, brings this to immediate pictorial visibility. The photographs and drawings incorporated here include examples of *living* water from a spring and of *dead* water from a big city. (See photographs I and X.)

Now what characterizes living, high quality water such as fully earns the description *wholesome?* We are accustomed nowadays to speak of the substance water as a chemical compound consisting of hydrogen and oxygen (H_2O), and to make only material use of it in technology. It would scarcely be possible for a hydrologist of our time to acknowledge the existence of any other attributes of water than those that can be summed up numerically. He reckons with the potential energy stored in a reservoir, or with the kinetic energy produced by water-powered turbines. Even biologists, those scientists of the life-element, work only with measurable and weighable numerical factors. The world of imponderables remains hidden from this kind of investigation, and these include that invisible world that imprints itself on water's *bodily* aspect.

Lacking insight into this realm, the science of hydrology knows only one half of the picture. And the other half lost to modern consciousness is just that portion that constitutes the secret of the life-element in water. We direct attention here to the salutary impulse growing out of anthroposophical spiritual science, which has provided descriptions of a world of forces on which the existence of every living thing is based. Without this insight, life's quintessence would be incomprehensible.

Attention is thus called to the real realm of forces, of etheric formative forces, raying in dynamically from the surrounding cosmos through all the rhythms that play between the heavens and the earth. That is why every genuine life-process runs its course in rhythmic patterns.

Water possesses a truly universal ability to lend itself to rhythms of every kind and to be active in them. Thus, it is able to develop in all organisms the rhythms essential to their life. Therefore, water is fit to become the bearer and mediator of the formative forces of the cosmos in every living thing, for the world's organic life originates not in the inorganic-physical realm but in spheres of cosmic forces. That explains the increasingly frequent finding that distinctly cosmic rhythms are to be found reflected in the life-processes of organisms. Research along these lines has opened up new perspectives in the sciences.

Much obviously still remains to be investigated in this field, especially where it is a matter of understanding water's total being. The challenge is to explore the part it plays as fresh, so to speak, virginal spring or rain water in the cosmic formative forces. Only when we have such information will we be entitled to speak with scientific justification of *living* water.

But a beginning has been made. It can be experimentally demonstrated by a new method, the drop-picture method, that conditions in the cosmic constellations find reflection in moving water, and what form they take there. Flowing wholesome water can be shown to take up the activity of cosmic forces, which become visible in changes in fine inner movements of water in accordance with conditions in the starry worlds. These inner movements mirror the changing quality of water under the influence of cosmic rhythms. The drop-picture method works with moving water and thus makes it possible to exhibit such movements in photographic form. It has provided insight into the way the cosmos imprints itself on water.[2]

The opposite also holds true, however. If, on the one hand, water quality expresses itself in water's motion, so, on the other, does corresponding motion affect the quality of water. If we fail to reawaken insight into water's cosmic aspect on the one hand, and to demonstrate and teach it, and, on the other hand, fail to take it into account and bring it to practical realization in the way drinking water is created and produced, there can be little hope for the desired improvement of quality in the water situation. The wisdom that once prevailed can be rediscovered on the basis of new insight, possible today, and made to flow again into the practical life of society.

Water Sustains All

Wolfram Schwenk

The theme "Water Sustains All," is enunciated by the Greek philosopher Thales of Miletus.[1] The quotation is taken from Goethe's play *Faust,* Part 2, Scene 2, which was written approximately 150 years ago.[2] It refers, of course, to clean, unpolluted water, for water pollution was rare in those days and, when it was discovered, it was sternly treated. But the situation has turned upside down in the 150 years since Goethe's time. Today, the water particularly in industrial areas is ruined; good clean water is the exception.

How far, then, does our theme apply today, and what reservations must be made in stating it?

Allow me, this evening, to refrain from a discussion of *everything* that water sustains. I will not, for example, go further into what is superficially viewed as the basis of our existence, namely, our economic system and its industries, which actually depends upon being allowed to pollute the water supplies. If industries were prohibited from doing so from one day to the next, they would have to shut up shop immediately. In other words, the life of the economy sustains itself on bad, polluted water. This is a high price to pay with life itself at stake as we shall see.

A lecture delivered on October 21, 1976 opening the Congress for Ecological Water Protection at Bad Oeynhausen, West Germany.

Let us, then, turn our attention rather to the indispensable basis of existence, to the planet earth with its atmosphere, to its soil, to its many manifestations of life represented by plants, animals and human beings. In these cases, the opposite of what was said above is true. They die if nothing but polluted water is available to them; their lives are sustained by good water only, not by water of just any kind.

Since industry—superficially held to be the basis of existence—operates sharply counter to what is really more vital to it, we must reach a decision as to how to bring industry into line with the vital basis of existence. A group of people has come here today to do that, and I am hoping to be able to provide you with some material that may clear the way to and ease that decision.

There is a second reason for our coming together here: it is the fiftieth anniversary of the drilling of the Jordan Spring, the earth's largest thermal source of carbonated saline water. It was drilled fifty years ago by Jordan, a very able engineer. The name was also borrowed from the River Jordan, to which the soil of Palestine owes a fertility it would not otherwise possess.

I have chosen to sketch the Jordan River as representative of the effects that water's activities produce. There, in the Holy Land, in the middle of a desert region, we find the sources of the Jordan and its tributaries welling up out of the earth and pouring an instant wealth of life throughout the area. There could hardly be a more perfect picture of the connection between life and water. The waters of the Jordan were the formative agency that shaped the earth in that region in prehistoric times. The soils of the lands bordering the river became fertile. We see here how closely related the earth's soil formations are to the watery element. The life-hostile climate of this part of the world, around the

Dead Sea for example, became bearable and life-supporting in the neighborhood of the Jordan and offered the population there the means of existing. All along the Jordan's course we find a belt of living green, surrounded by infertile desert country hostile to life. Thus, water shows itself to be the medium whereby deadly terrestrial extremes are brought into balance and, in the resultant synthesis, made to serve life.[3]

If we now turn our attention to the water incorporated in living organisms, we discover it to be the indispensable structuring medium in every life-process and in every organ. It is the most important constituent of all the bodily substances of which plants, animals, and human beings are composed.

For example, proteins owe all the structural stability in the arrangement of their chemical building blocks, and thus their functional capacity, to water. Water, in the form of so-called structural water, makes up the major part of the volume of these proteins, thereby serving as the sole guarantor of their existence.[4]

Water, as we ordinarily know it, is the most formless substance imaginable. It slips through our hands as we try to grasp it. It streams down a brook adhering to every shape it encounters. Completely renouncing any form of its own, it contributes in living organisms to the support and stabilization of structure in proteins, those most vital bodily substances. What amazingly contrasting activities it carries out!

Let us take a further example from a higher realm of life, the human body. We find that a newborn baby's body is made up of over 90 percent fluids, while an adult's runs to approximately 60 percent. But how little these properties convey until we call to mind how water moves and what transformations it undergoes! Adults ingest an average of about two liters of water daily with their food. These two liters comprise about one-ninth of our daily extracellular bodily fluid exchange. That is an impressive amount, even though it is approximately only one-thirty-fifth of our body weight.

Babies, on the other hand, take in about one liter daily, replacing about one-third of their total extracellular body fluids. This approximates about one-sixth of their weight.[5] Thus, in a week's time, babies convert more than their entire weight in the form of water—a most lively flow of water through their bodies.

Water has many different functions in the body. It is the most important metabolic agent, responsible for breaking down and expediting foods for conversion into bodily substances. It shares the work of incorporating, structuring, and depositing these substances, and it is the transporting medium in their conversion, breakdown, and elimination. The entire metabolic process is bound up with water, in a watery milieu. Furthermore, water makes itself responsible for regulating body temperature; human beings need close to one-quarter of their basal metabolism, the energy used up by the resting body, for the evaporation of water from the body's surface just to keep the temperature stable.[6]

To sum up, we find that water—observed externally in its fluid state—is completely formless; that it adheres to and takes on the shape of any object with which it comes in contact; and that it flows downhill in response to gravity. On the other hand, we find it selflessly helping to support structure in protein building blocks and adapting itself to the structural pattern of protein molecules in order to stabilize their highly specific structures. Water is colorless to the observing eye; good water gives the impression of being tasteless and chemically neutral, that is, without chemical imbalance; and we perceive it as restless, as in constant motion. This means that we have to characterize water as a substance of purely negative attributes, tending to the end-syllable "less,"—as, for example, in restless, tasteless, colorless, formless, and selfless.[7] And life is immutably bound up with this selfless water.

We note from a brief review of its physical attributes that this selfless substance water does not fit into the periodic tables of matter; it is "unconventional." It is conspicuous for departing from the norm in that it does not invariably increase in density upon being subjected to increased chilling; instead, it reaches the point of greatest density at 4°C, growing lighter thereafter upon undergoing further cooling and turning to ice. This has far-reaching consequences for the configuration of the earth's surface and for the way life is distributed upon it. I merely remind you of this fact without going further into the matter here.

Water departs from the norm in another aspect: its melting and boiling points. If water conformed to the way chemically related substances behave, it would melt at circa -100°C, and reach the boiling point at -91°C. However, it actually melts at 0° Centigrade and boils at 100°C. This anomaly is due to its exceptional structure, which is based on the molecular structure of water and accounts for the fact that water remains fluid within our temperature range. This is thought to result from an infinite variety of interconnections of the molecules of water and higher molecular aggregates.

Lastly, water also departs from the norm with respect to its least specific warmth which reaches a minimum at 37°C. This is just the place in the temperature range where water's absorption and loss of warmth are particularly favorable for living organisms,[8] and exactly that which constitutes the normal body temperature of human beings—the most highly developed form of life.

Thus, the overall picture derived from these facts conveys the impression that water does not fit into the framework of the normally prevailing laws of matter. Water evades inclusion in the periodic table of the elements. Yet life is particularly closely linked with this evasive substance.[9] We are therefore justified in saying that life depends upon flouting the laws governing inorganic matter for its existence on earth.

But water eludes our grasp again and again. Whether you view its molecular structure as a bent lever with the oxygen atom at the center surrounded by the two hydrogen atoms, or look at its various physical measurements, conflicting statements are still altogether possible.

One of the most recent international symposia on the structure of water and aqueous solutions gave a classic demonstration of how the exceedingly simple basic structure of water opens the way to the most conflicting approaches, and of what riddles professionals in the field of physical chemistry face.[10] Despite their exact measurements they still cannot say whether this or that structural formula is correct. The conclusion could be drawn—something physical chemists themselves do not do—that water possesses a tremendous range of freedom, alternating in the most varied interplay and making it possible for the laws of life to take effect.

Despite our review of the dependence of life upon the presence of that most unique fluid, water, we must agree that it throws no light on water's sustaining functions. What, actually, does "sustaining" mean? In the case of lifeless objects—works of art, for example—it certainly means "preserving" them or keeping them as they are. But this definition is just as certainly inapplicable to living organisms. If a living creature were to be left as it is, it would quickly die. Here, "preserving" or "sustaining" means rather to keep a living organism functioning, to keep renewing it, not in the sense of producing something different from the old, but supporting it in an ongoing development. If we now include this meaning in the title of this evening's talk, it will have to be amended to: "Water sustains renewingly all life, while assuring continuity of the organism." How is this to be understood? What possibilities are at hand for pursuing such an investigation?

"Preserving," "sustaining," "renewing." You are surely familiar with the way fairy tales, myths, and sagas tell of "the water of life." Wherever forces of renewal play a role in these stories, water is their symbol. But it is also the element used to picture the transporting of man as an earthling over into another world—the element that mediates between the earthly and the spiritual realms. Think of the story of Mother Holle, in which a child descends into a well to enter another world and then ascends out of it upon returning to earth. Remember the wide waters in myths and sagas that had to be crossed in order to leave one world and enter another. People of earlier times obviously experienced physical and spiritual renewal in relation to water while still maintaining their individuality.

How can one achieve this afresh in our time on the basis and with the strength of real insight? How can it be made accessible to our understanding?

Let me also mention something else. In olden times people encountered, here and there, spring-gods and river-gods in their experience of water. We may at least conclude from this that they did not regard water as merely H_2O plus a few other properties, as an element everywhere identical, but rather as possessing individuality. We are justified in thinking and speaking of bodies of water in the same way today. If you talk with hydrologists or examine the literature on limnology or oceanography, you will scarcely find a single professional making generally valid statements about all waters of the same type. Whatever is stated applies purely individually to this one spring or that one river, to this particular sea or to that ocean region. This is recognition, no matter how dry and superficial, of the fact that the waters of our earth are indeed individuals.

What makes water the element that brings renewal to living organisms while maintaining continuity?

Certain clues can be gleaned from what we have been considering today. We have, for example, discussed the intensity with which fluids stream through babies' bodies. In external nature, water flows downhill. It responds to gravity, collects behind dams, then overflows and streams on again. But in living organisms water circulates and rises. It circulates and rises in the atmosphere as well, but there it vaporizes, and changes into a different aggregate state. In organisms, it rises always in the fluid state, and in human beings it comes to rest only upon reaching the highest point, the head. The brain does not respond to the downward pull of gravity and locate in the legs or metabolic region; it is at the very top of the organism. That is the point to which the water in the body has risen,and the brain, afloat on this buoyant fluid, is thereby enabled to serve as the physiological foundation for the activity of thinking.[11]

All these illustrations point to the way water moves. This raises the question of what water is doing in its streaming motion. Are there laws hidden here whose discovery could bring us closer to the problem? Let me remind you that all the various life-serving activities of water carry out their functions only when water is in motion. The water that has come to rest in a chemist's or physicist's test tube cannot support life-processes.

Water capable of doing that is moving water. So now let us consider water's motion—with the help of the findings set forth in Theodor Schwenk's book, *Sensitive Chaos*.[12]

Even when the water in a waterfall encounters no obstacles in its response to gravity, differences show up. As it trickles over a projecting cliff, it forms single threads rather than an unbroken sheet. A brook describes a meandering course in its downhill run, swinging rhythmically from side to side. It is a stranger to the straight, shortest route. Obstacles lying along its banks or on its bottom interfere with the free movement of its water. But instead of slowing its pace while retaining the same form as it passes over them, it forms waves that remain in place and are overlaid by smaller and capillary waves. Thus, rhythmical patterns are brought into being. New water is constantly flowing through these stationary waveforms, exhibiting a phenomenon basic to all forms of life: a retaining of form combined with an exchange of matter, and a constant renewal of form as new matter flows through it, while continuity is preserved.

Wave-play of this kind stirs up sediment on the bottom of a stream and arranges it in rhythmical patterns that demonstrate water's liking for movement on intertwining planes. Depending on conditions, patterns like the channel-systems seen in shoal waters that remind us of plant forms, or, in their serial arrangement, of fish skeletons, come into being. Then again, currents flowing in intertwining layers call forth patterns that recall the structure of organs, or the way muscles lie. Regardless of whether such currents are observed in nature or experimentally produced, there appear momentarily in flowing water indications of patterns reminiscent of familiar organ structures, throwing new light on organ morphogenesis. Inducing trains of vortices in viscous fluids creates models that demonstrate how structures reminiscent of the leaf buds of plants are produced by strata streaming past each other at varying rates of speed, and how curled and matted structures reminiscent of the root realm originate where environmental factors are particularly obstructive. Where there is a more intensive clash between obstacles and movement in a fluid, whole trains of vortices come into being, with alternating pairs of vortices curling strongly inward and forming hollow spaces. These extensive trains of vortices occur in living nature especially where overwhelming obstacles obstruct growth movement, as, for example, in the wood of trees growing in high mountain regions and in the bony structure of a deer's organ of smell, which is filled with a mass of bony trains of vortices.

Bony structures may even reflect such subtle details as the way the absolute lines of a vortex flow continue over into the surrounding medium, past its boundary and at right angles to it, as in the arrangement of bony trabeculae composing and supporting the structure of the bones. In the joints, these continue similarly from the ball to the socket, and at right angles to the surfaces. The inwinding and curling movements of such paired trains of vortices follow the same laws as do the involutions seen in animal embryos and organs in the early stages of their development and right down into the subtlest detail.

How are such trains of vortices produced? Surprisingly simply—by just drawing an evenly moving object in a straight line through a still fluid. If the hydraulic conditions are right, the fluid does not respond—as the study of physics would lead us to expect—by approaching the most probable condition: maximal chaos. Instead, it forms vortices related to each other in rhythmical sequences and arrangements. There is no rigid mechanical repetition, but rather a rhythmical sequence of vortices, similar but not identical to one another, so that the beholder is treated to a display of ever-changing vortex patterns. Thus, mutually related structures come into being, giving rise, in the streaming water, to patterns quite concretely related to those out of which the forms of organs are developed.

One is entitled to speak here of "organizing" currents, in the most literal meaning of that term. We should note for scientists that the processes described take place with a lessening of entropy.

Trains of vortices are also produced whenever a continuous flow of water is directed into still water. The resulting series of paired vortices can serve as an actual model of the morphogenesis of an insect's heart, based as it is on a serial arrangement. To visualize this open-sided organ, we have only to picture the hemo-lymph flow enveloping the train of vortices it has brought into being. But the hearts of other more highly organized living creatures, including that of human beings, also demonstrate the laws of flowing motion in their forms; the muscle fibers of the left chamber of the human heart with its interlaced double vortices clearly demonstrate their relationship to the movements in pulsating vortex funnels. The double vortex in the arrangement of muscles at the apex of the human heart is a further example of the same principle. We can see from such a contemplation of the way form is born of pulsing movement how, at every heartbeat, the heart form is freshly created out of the streaming blood, how the renewing of the heart's form is provided for with every heartbeat, and how continuity is created.

And what has been demonstrated as applying to the

heart can be developed quite similarly for the lung form, with the production from simple vortex-currents of the trachea, the bronchi, the two lobes of the lungs, and the lung-sheath enclosing the whole. Here, too, we can arrive at the concept of this lung form being given shape and continuity with every breath. And the other organs—ears, intestines, and so on—can be understood as both originating from and being maintained in their functioning by movement. Our theme "Water Sustains All," while assuring continuity, has now taken on very concrete aspects.

No matter how differentiated these organizing currents may be, they are all interrelated; this means that they move in close coordination. Let us take the example of a ring-vortex as it moves through the various stages of its development. Its rotation-axis is a closed circle. It can be produced in water, and by expert smokers in the air.[13] If, on encountering an obstacle, it is forced to unroll and turn over, extremely striking star-shaped patterns of related vortices are formed, which again display a tendency to a great variety of forms such as appear in the development of organs.

To sum up, we find that water in motion actually combines every organ-shaping possibility in a universal, as yet unfixed form, and transmits these to living organisms.[14]

Experience with other flowing media shows that organizing currents are not tied to just one particular substance. The creation of organ-like forms by flowing water cannot therefore be attributed to water as a substance; it comes from a quite different realm. We can draw on the most recent expert opinion for support in this view by citing Professors Wilkinson and Bordner of Philadelphia. They say that the earth is a gravitational body in the cosmos that holds life together, while the cosmos is the universal regulator of life-functions.[15] And it is just in water, strangely enough, that cosmic events are reflected.

We have found in our own research that movement produced in water under identical experimental conditions varies markedly according to the time and the cosmic situation prevailing at the moment. We find that there are moments when the readiness of water to produce organ-like forms is enhanced, and other moments when it is so inhibited as to produce caricatures.[16] And we also find that these moments are not haphazard, but quite concretely correlated with astronomical events such as the aspects of planets, for example, when these form certain particular angles, dividing the circle of the sky into whole number relationships, or when there are eclipses or solar eruptions, and so on.[17] The research we refer to is far from complete. But it has advanced to a point where we now know that water is not just an element in movement here on earth, but is instead fitted—perhaps just because it breaks free from and transcends earthly laws in so many respects—to serve as an instrument and medium for those life-directing events brought about on the earth by cosmic influences. A wealth of indications on this score is found in the lifework of Rudolf Steiner, on which, in turn, our own work is founded.[18]

"Flow patterns are always motion patterns in which all the forces involved—and this includes imponderable forces—are in balance."[19] It is therefore easy to tell from the forms perceived in fluids whether they are in a state of balance, of inner equilibrium, with respect to their constituent factors and components. Good water is found to be in just such a harmonious and balanced state. The flow patterns it produces are therefore always organ-like. When we pollute water, we bring imbalance into it and perceptibly disturb the harmonious interplay of the factors and components of which it consists. The result is that such water can no longer move in organ-like patterns, which is to say that it is no longer receptive to life-governing cosmic forces and, hence, cannot transmit them to living organisms. This is impressively demonstrated in a special experimental procedure called the "drop-picture method."[20] A flow-process is induced in the water to be tested by letting drops of distilled water fall back into the specimen and the differentiated patterns are rendered visible by means of the schlieren-optical method. The luxuriant and varied possibilities of form-creation inherent in fresh springwater are completely destroyed by the incorporation of waste water, so that almost formless, undifferentiated patterns are all that remain. (Photographs III and IV) Nature eventually succeeds in restoring balance in moderately polluted waters by means of biological self-cleansing processes, and this is reflected in stepwise changes in the flow-patterns, but in this country we have already gone beyond the limits.

When we human beings rob water of the possibility of organic motion, we make it a stranger to its life-serving function and destroy its capacity to achieve renewal. When we think of water as a dead substance and treat it as a mere carrier, or as just a source of energy, we take away its life-sustaining power.

The decision as to whether all life can, in future, be sustained by water and continue to experience renewal lies in human hands. It involves understanding that water can tolerate only limited burdens of pollutants if it is to continue in its life-sustaining functions. So the question of good, life-sustaining water becomes, in the last analysis, a question of the state of mind of every human being. Nothing is accomplished by accusing some industry of polluting this or that body of water; industry produces only what we demand of it, for we want to be able to pick up goods at the nearest shop. It is ours to decide whether, by establishing consumer habits beneficial to water, we will begin to set an example that will gradually change the consumer habits of the population at large and lead to a different emphasis in production.

However, this is not the place for me to hand out a recipe and say: "Do this, or do that!" I saw it rather as my task this evening to make a contribution toward a renewal of awareness of the laws of life in the watery element as the prerequisite for a modern water-consciousness that might serve as the foundation for the right way of treating our life-source—water. Then it is your task, everyone's task, the task of all of us as a community, to decide in freedom and insight what future course to follow.

DROP PICTURES

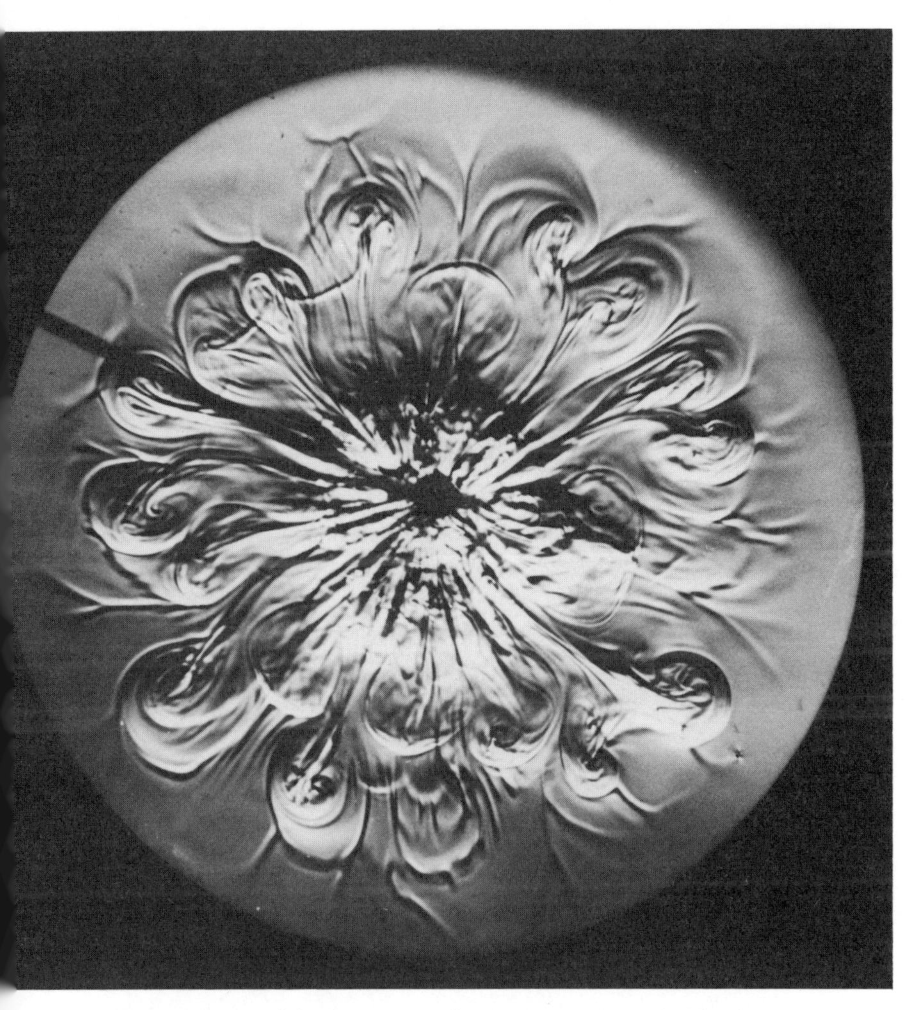

I. Drop-picture of drinking water from a mountain spring issuing from the lower triassic quartz sandstone formation. "Living" water. Manifold, pronounced leaf-shaped vortices are coordinated here in a rhythmically structured, harmonious rosette pattern.

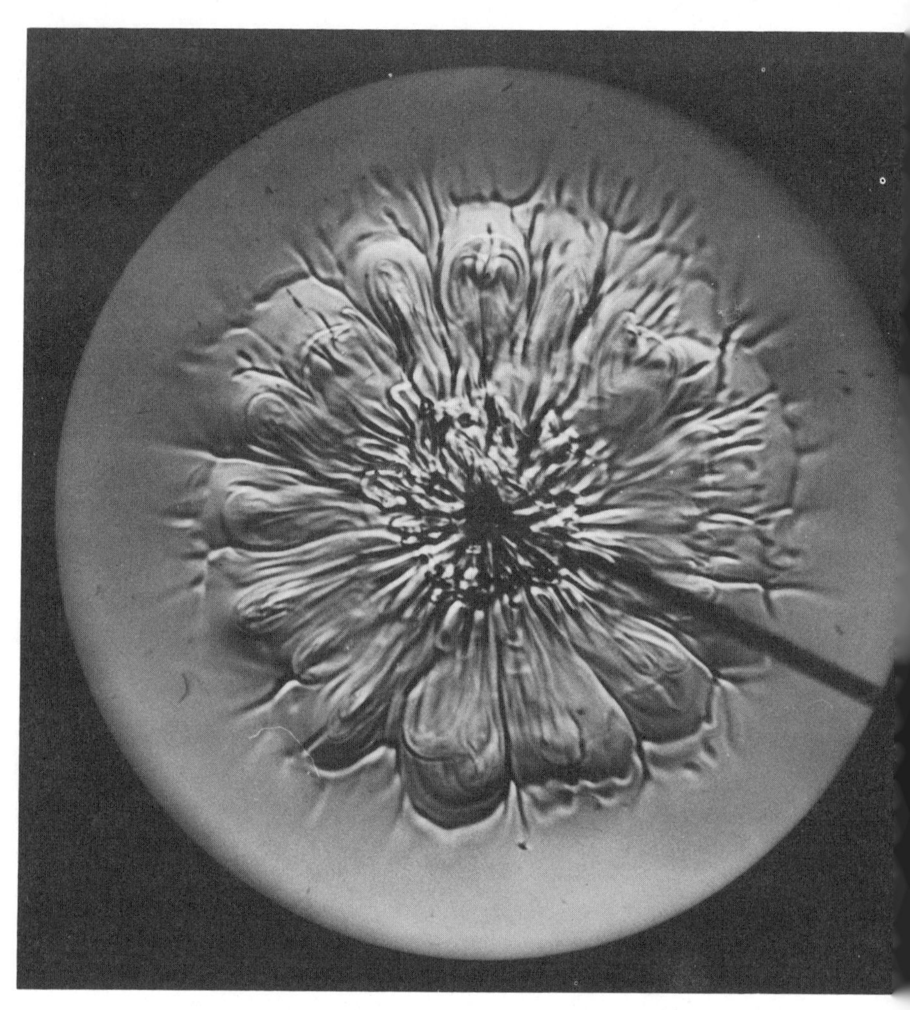

II. Drop-picture of the hygienically acceptable water of a large industrial city in West Germany, where groundwater is mixed with recycled water. Though this water is unobjectionable from the hygienic standpoint, residents of the area feel it to be "bad" and "dead." The rosette pattern is poorly developed. The leaves of the vortices are merely rudimentary and melting into one another.

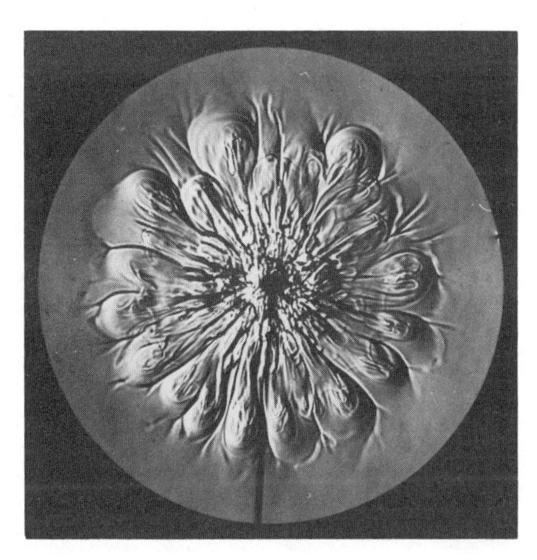

III. Drop-picture of water from a mountain brook in the Black Forest, taken from a point below the source. Living, naturally flowing water. Drinking water quality. The rosette is well developed, with a broad spread of vortex leaves. Example of a less lively modification of the normal type.

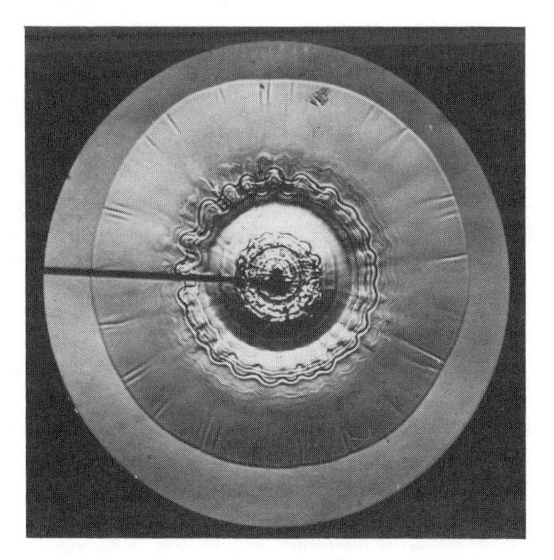

IV. Drop-picture of a water sample from the same brook from which the water in picture III was taken, but in this case from a point further downstream after the entrance of domestic sewage and industrial effluents. The disk shows a mere trace of rudimentary development, with little differentiation. The formative capacity of this water is extinct.

V. Drop-picture experiment using spring water. Photo taken at third drop.

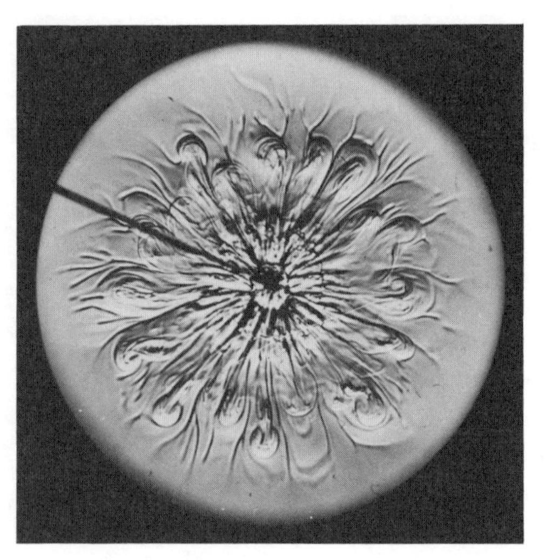

VI. Photo taken at twentieth drop.

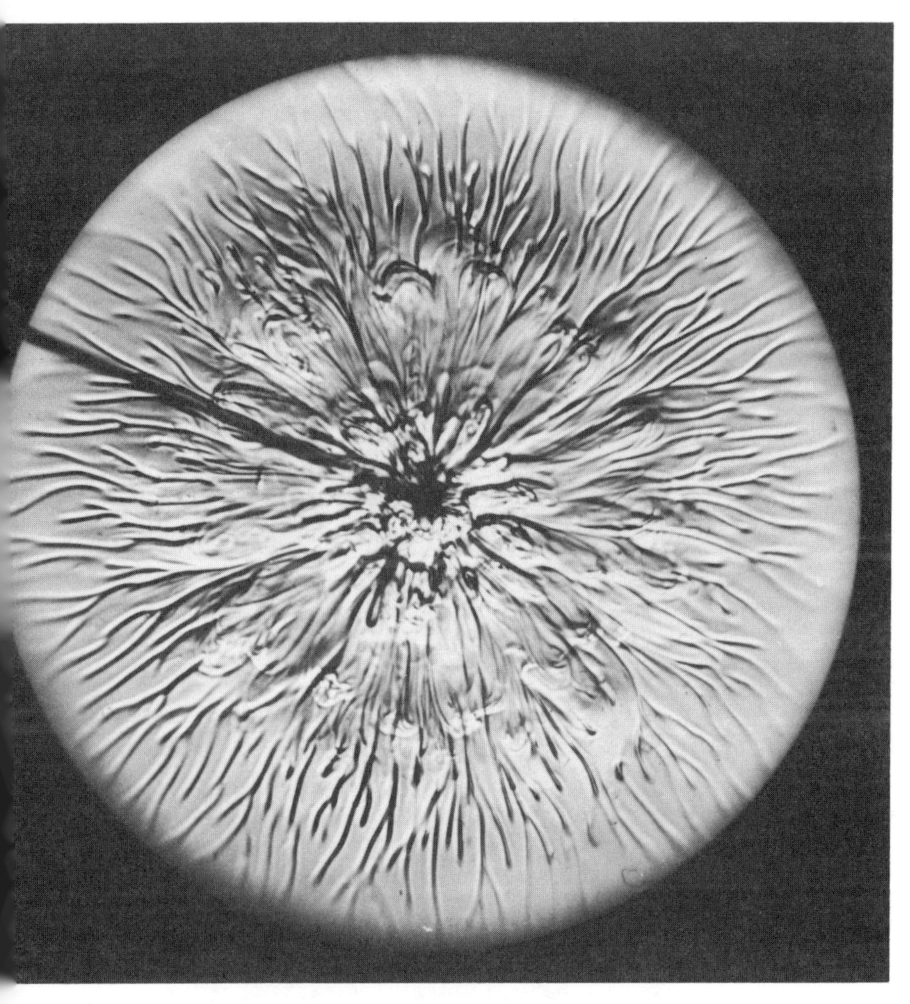

VII. Photo taken at thirtieth drop.

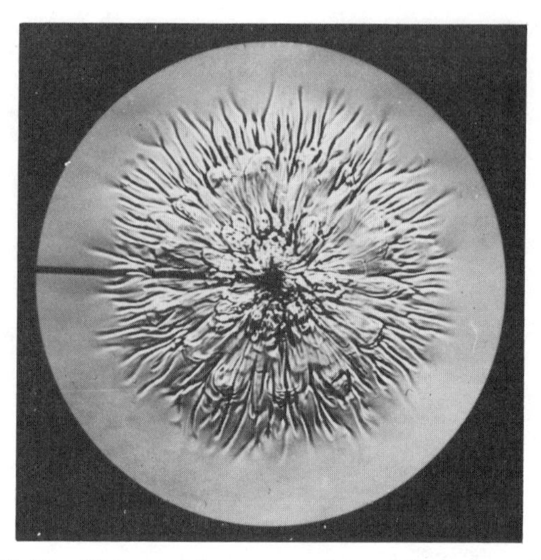

VIII. The Rhine surface water about 40 km above Basel.

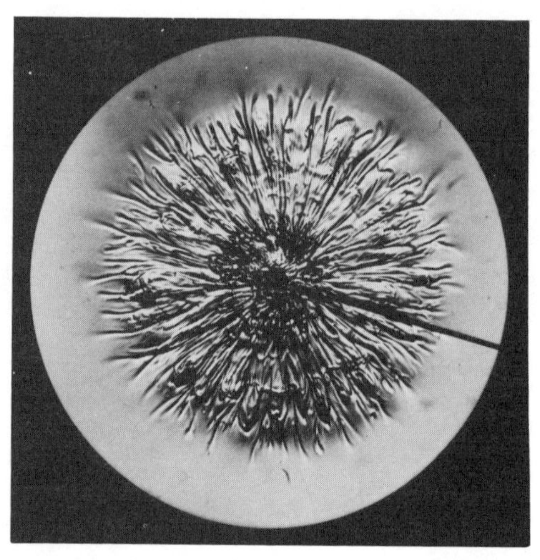

IX. Big-city drinking water, processed from filtered water of the Lower Rhine river.

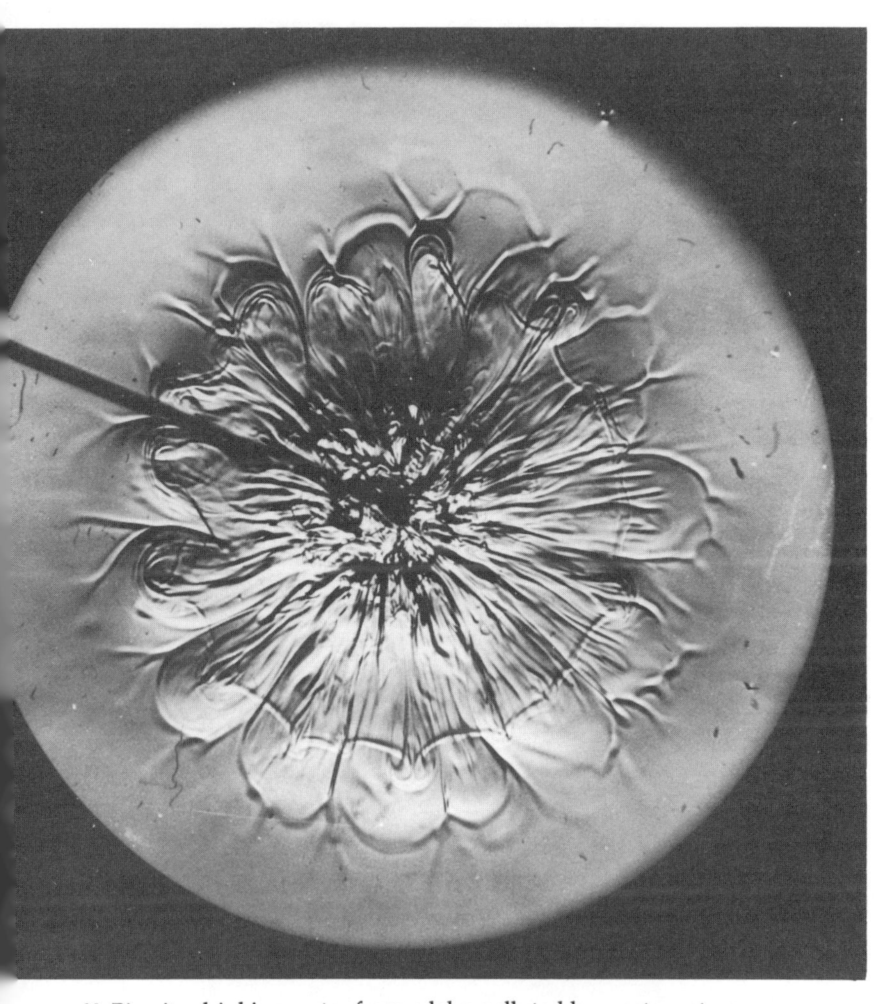

X. Big-city drinking water from a lake polluted by waste waters.

XI. Medicinal water for intestinal illness after the twenty-fifth drop.

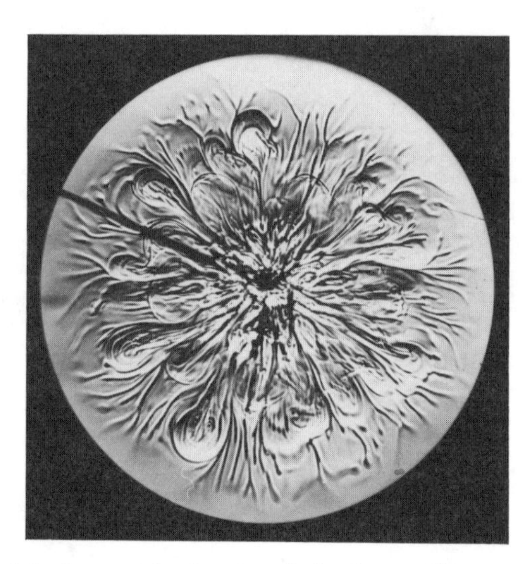

XII. Medicinal water for heart and circulatory illnesses after the twenty-fifth drop.

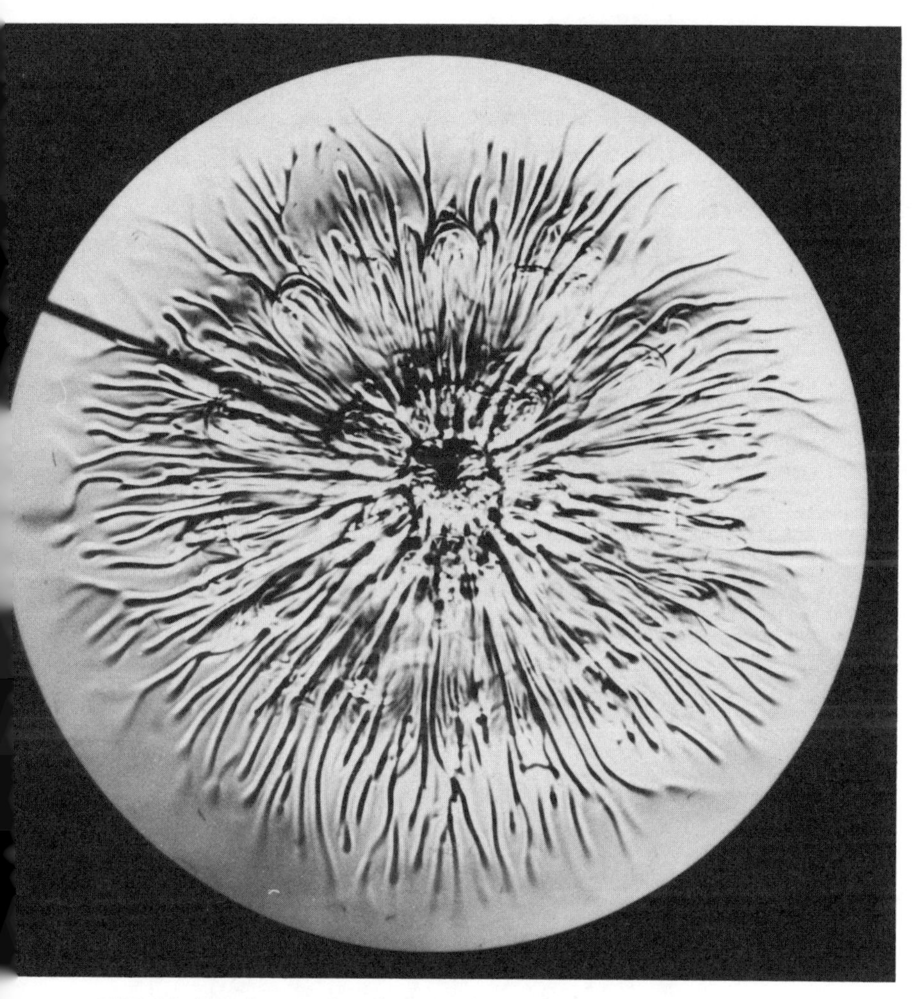

XIII. Medicinal water for ailments of the skin and nerves after the twenty-fifth drop.

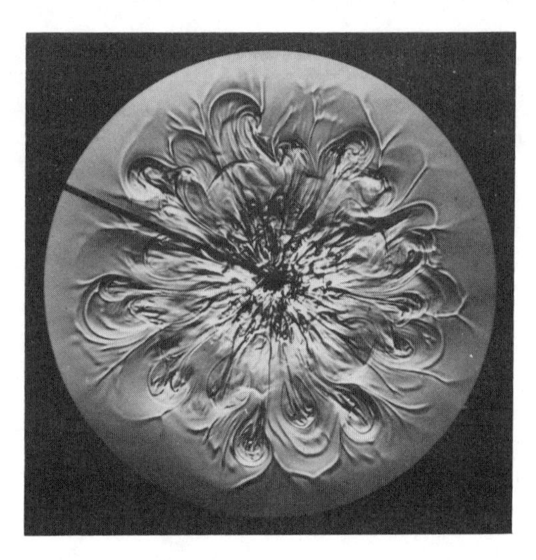

XIV. Drop-picture of living water from a granite spring in the Black Forest. The rosette is well and evenly formed throughout and has many richly varied vortex forms. This is an example of a very lively modification of the normal type.

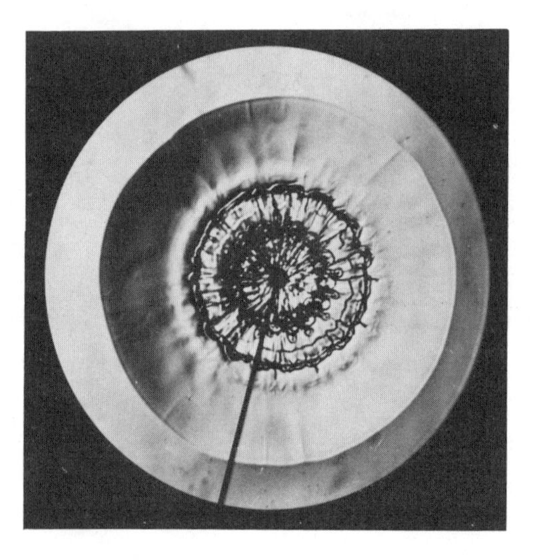

XV. Drinking water of a big city, processed from the water of a detergent-polluted lake.

Testing for Water Quality: The Drop-Picture Method

Wolfram Schwenk

Drinking water supplied by municipal water departments comes to the consumer in a condition that, from the hygienic standpoint, is considered unexceptionable, and hygienists consequently rank it as "good." Despite this scientific dictum, however, the consumer often feels that the quality of water thus approved leaves something to be desired. A study of the latest statistics compiled by the Association of German Mineral Springs indicates how prevalent that feeling is, for it appears that more than ten times as much bottled water is now being drunk than was the case thirty years ago. Since about 1966, "the sales curve has lost its previous seasonal peaks, due to a change in drinking habits, and it has remained level." Two-thirds of this bottled water is being supplied to private households.

What is the difference between the ways hygienists and consumers judge water quality?

The water hygienists base their judgments on analysis, and if they find no chemical or bacterial adulteration in a water specimen they call it "unexceptionable" or "good." The absence of a few selected negative factors allows them to make an overall finding of satisfactory quality. Their procedure in this resembles analyzing a text by looking for and sorting it by certain letters, a

Reprinted from *Das Seminar,* 2/80, pp. 39–51.

process in which the text remains unread. But alert consumers learn more about water quality from their sense impressions than they do from lists of deleted components; it is not so much this or that single aspect as it is the character of the total fluid, water, that their senses register. And this character is the product of the way all the components and factors involved work together rather than merely a question of which of these is present or absent. Thus, consumers unconsciously practice a synthesizing form of judgment. To use our metaphor again, they perceive the text, in other words, they read what originates from a combining of various factors and components, giving scant attention to what letters are present, and to what extent, in the words and sentences.

Both ways of judging are one-sided; they stand in need of mutual complementing. But where the consumer readily accepts this complementation from the hygienists' side, the reverse is not yet the case. Hygienists and analysts do not recognize the validity of the different principle on which consumers base their qualitative approach; they see it as something that cannot be scientifically measured, and therefore reject it entirely as subjective. Now the uniqueness of an entity that is the product of the way its components and factors work together cannot be characterized by a mere measuring and enumerating of the individual elements of which it is composed. But this fact cannot be held to prove that such an entity does not exist.

The need is rather for the development of an appropriate method whereby uniqueness of this kind can be grasped and that enables us to apprehend such an entity by allowing it to demonstrate its wholeness instead of trying to present it in the form of a breakdown of its parts. So the method chosen here will be that of observing and demonstrating the behavior or character of the object under study rather than of its components.

Theodor Schwenk's drop-picture method is an attempt to apply a procedure of this kind to water, to grasp its behavior as the product of the interaction of its components and factors, and to provide supplementary information in order to add to a description of the qualities of this fluid.[1] Since the nature of water expresses itself particularly clearly in its flow-forms, it was decided in applying the method to be described here to work on the basis of the subtlest differentiations in flow patterns.

The Reynolds Equation for similarity in flow-motion can serve in first approximation as the rheological (physics of flow behavior) basis for this study. According to this equation, similar flow patterns can be produced under varying conditions if the product of certain proportions of material factors of the flowing liquid is constant with the external conditions of flow. On the other hand, one must, according to this criterion, reckon with different flow patterns in two liquids in cases where their material factors stand in different relationship and the external instrumental flow-conditions are kept constant. The flow pattern can then be made to reflect the different proportions of the material components in the flowing media. If the flow conditions for a particular fluid are then adjusted to the labile border region between the turbulent and laminary flow areas, very subtle shifts in the balance of factors can be read from the way forms develop in the current. In the drop-picture method, the parameters of the apparatus used are so meticulously adjusted and balanced that a highly sensitive flow reaction results and can be pictorially rendered.

This experimental technique is based upon the following principle: The water specimen under study is contained as a thin layer in a level, round glass bowl, and is then brought into motion by a number of drops of distilled water falling into it at regular intervals, in the course of which currents are set in motion. A homogeneous solution of the water sample, to which glycerin has been added, is used as a neutral, inert schlieren agent. The patterns formed are rendered visible by means of a schlierenoptical device adapted from Töpler by Theodor Schwenk, and these are photographically recorded. For best results, the calibration of the drop-forming apparatus is performed with steam-distilled water of unvarying quality available for use.

The flow pattern that comes into being in the sample after a drop has fallen into it remains almost unchanged for the few seconds before the next drop enters. The pattern changes somewhat with the entrance of every further drop, so that the characteristic changes are continually occurring as the experiment runs its course from the first to the thirtieth drop. Single delicate vortices radiate in all directions from the center of the bowl where the drops strike; they form a rosette that develops from drop to drop as the leaflike vortices spread out, multiply, and undergo differentiation. An intensification process takes place in this development up to the twelfth drop, ebbing again after the twentieth drop, at which point new, rodlike structures appear in radial arrangement and increase to such a degree that at the twenty-fifth to thirtieth drop a completely radiating type of pattern makes its appearance. If still further drops are added, this radiating pattern begins to disappear. A poorly articulated, ring-shaped zone pushes into the space between the periphery and the center where the drops strike. This expands gradually until it becomes an almost undifferentiated disk or sheet, and ends at last in a dissolution of the previous structuring.

When various other water samples are brought into motion under identical experimental conditions and subjected to the developmental course described above in its application to a spring water specimen, a greater or lesser degree of modification results. In addition to the differentiated shaping of the single structural elements, it is particularly striking that the transition from the leaflike rosette type of pattern to the radiating type can occur at an earlier or later drop-count, depending on the nature of the specimen. It therefore usually suffices, for a first orienting comparison of specimens, to compare their flow-forms at a predetermined drop-count, say the twentieth, as was done in the case of the samples in the illustrations. (Compare photos III–IX and XIV and XV.) For instance, surface waters, in which qualitative differences can be clearly demonstrated by means of traditional testing methods, also show basic differences in the manner of their flowing.

Thus, the quality of water can come to expression in the way it flows; its quality and motion are related. And as this relationship is studied, the determining of quality is not limited to selected, checked parameters; it covers the interaction of all the various components and factors, reflected in movement.

This procedure demonstrates that hygienically unobjectionable drinking waters can span just as broad a spectrum of types of motion as do surface waters ranging from refreshing springs to brooks freshly laden with domestic and commercial waste water. Such comparisons bring home to us how inadequate descriptions of drinking water quality must remain if all they do is supply data derived from analytical measurement, for very considerable differences between hygienically acceptable waters can show up in objective and reproducible form in their behavior. Movement patterns found in drop-pictures of the drinking water of various big European cities display a wide panorama made up of all the levels encompassed by the span between the aforementioned polarities.

Conversely, flow patterns produced by natural spring water samples resemble one another so closely that they can be allocated to a common type of pattern, even when the springs they are derived from are of very different hydrogeological origin and hydrochemical make-up. At the twentieth drop, this type produces a distinctly formed perfect rosette, consisting of a leaf-shaped arrangement of vortices that have not yet coalesced and between which rod-shaped elements move toward the periphery, though at this stage this pattern is not predominant. There is a variation in this type from more to less lively and from more to less luxuriant patterns.

This pattern common to spring water specimens serves as an archetype for the flow-forms of drinking waters sensed as having a good, refreshing, and revitalizing effect.[2] Thus, the drop-picture method meets one of the basic requirements made by *DIN 2000 (DIN* = German Industrial Standard), succeeding where hygienic and chemical standards have thus far proven inadequate.[3] "The requirement for approval of drinking water has generally been that such water measure up to the attributes of ground water derived at a sufficient depth and from sufficiently filtered veins, water chemically acceptable, and from a self-renewing source that contains no trace of pollution."[4]

But this pattern is certainly not to be thought of as eliminating the necessity for hygienic testing; it simply supplements the latter in the sense described. For as the contrasting of drop-pictures of ground waters not always unobjectionable from the hygienic standpoint with samples of hygienically acceptable drinking water shows, flow-forms do not always conform to the hygienic findings. Similarly, tests of many samples of water of hydrochemically differing composition demonstrate that the drop-picture method provides no chemical-analytical information about drinking water, and can therefore never serve as a substitute for chemical analysis or render it superfluous. The strength of the drop-picture method lies rather in providing supplementary information about the similarity of a drinking water sample to fresh spring water in cases where the drinking water has been shown by reliable hygienic and chemical analyses to be unexceptionable. The contribution of the drop-picture method here is that of a supplementary indicator, supplying answers to questions which cannot, because of the difference in the principles involved, be answered by an analysis of chemical and physical factors.

In cities availing themselves of more than one source of water supply, the drop-picture method provides information as to which of the sources most resembles natural spring water, and its behavior in movement clearly distinguishes the latter type of water from specimens of processed surface water. With the outcome evident before one's very eyes, it is sensible to recommend that spring-derived drinking water not be mixed with processed water of lesser quality, but rather made available in public reservoirs to people who consider it a matter of importance to have access to a natural supply of drinking water and are therefore willing to go to such storage places to obtain it. Even though we know that the water supplies of many large cities no longer deserve to be ranked as "good" in that term's fuller meaning, we cannot share Dr. F. Morell's conclusion that the bottled water from only two springs remaining in all of Europe is fit to be recommended for drinking purposes.[5]

We think it would make more sense to demand that the good wells or springs located in every part of this country be kept for nutritive purposes (this comes to a mere 2% of potable water requirements) and made available. But this certainly does not mean requiring a second set of pipelines to be installed to bring this spring water into people's houses. When spring water stagnates in a pipeline, the loss of quality is so considerable that it becomes quite obvious in a drop-picture. Making natural spring water available must therefore mean that people go to the wells and springs to fetch water in the quantity needed to serve nutritive purposes. It can be gathered from the statistics quoted above that a large proportion of the population would willingly undertake this.

It would not be telling the whole story to describe deviations from the archetype in the flow-motion of various potable water specimens in terms of a mere set of concepts about a stepwise lessening of water quality. Our efforts at the Institute for Flow Sciences to establish a grading system for hygienically approved drinking water based on comparison with this archetype are therefore presently focused on learning to determine the significance of the various degrees of deviation from the archetype. Our attempts to gain experience with samples of water from mineral springs are to be seen as heading in this direction. Medical experience has shown that the waters of many such springs often have quite specific, physiologically one-sided effects on the users' organisms. Very strange parallels turn up when the various types of motion seen in drop-pictures of water specimens from these springs are compared with their therapeutic effect when ingested. I would like to describe them from the standpoint of our present understanding as follows: Mineral waters affecting glands and metabolic organs are reflected in drop-pictures in especially marked roundings, not only in the vortices that continue developing strongly and luxuriantly even at the twentieth drop and frequently push out a series of plateaus raised stepwise one above the other, but also in the rodlike structures, which, in these cases, incline toward each other and often coalesce in bow-formations (compare photos XI–XIII).

In contrast to these, drop-pictures of mineral waters with therapeutic effects on the nerve-sense organism exhibit a prematurely radiating type of pattern, a straight-lineal structural tendency. Whereas waters used for the therapeutic treatment of the human rhythmic system—heart, breathing, and circulation—and which therefore affect the physiological system that mediates between the poles of the nerve-sense system and the metabolism, also display a balance in the morphology of the corresponding drop-pictures, a balance between the exaggeratedly rounded forms of the one type of water and the one-sidedly radiating patterns of the other. There are, of course, very great individual variations within these overarching types, such as are also usually found characteristic of the different springs belonging to the same general therapeutic grouping. But it is especially striking that type-parallels between the flow-forms and the therapeutic application are also present in cases where the hydrochemical make-up of such waters does not suffice to explain their therapeutic effectiveness, and would therefore not lead us to expect any such correspondences.

Comparisons of the drop-pictures of various drinking

water and mineral spring water samples bring the one-sided characteristics of many big-city water supplies to actual visibility, whereas the spring water archetypal drop-picture shows a close resemblance to the mineral spring water type that brings about a balance between the nerve-sense and metabolic extremes and thus avoids one-sidedness. The archetype's usefulness as a model in gauging the universal nutrient status of a drinking water specimen is thus confirmed through the human organism, and the more or less one-sided flow-motion character of many drinking water supplies can also be determined in this way.

I will touch very briefly here on a theme that comes up in studying the differentiated ways water moves, for it involves factors that must be taken into account in performing experiments. Even in the case of good water specimens, deviations from the normal pattern can turn up in drop-pictures during the shaping of the delicate flow-forms. These deviations come about at times that can be exactly pinpointed and very often predicted. They are related to astronomical events.

Sun and moon eclipses, certain angles of relationship between the planets and other planetary phenomena, including their daily risings and settings and culminations as determined at the particular location where the experiment is performed, can be reflected at the moment they occur in changes in the delicate currents of a water's flowing.[6] Changes of this kind sometimes have very little effect on details, but on other occasions they can be so great as to result in a different type of pattern. For the carrying out of experiments, this means that such largely predictable moments should be avoided for quality determinations in cases where a water specimen's way of flowing is to be examined. This makes it essential to look ahead and plan the timing of experiments very carefully in working with the drop-picture method. Trainees in the method are taught to keep this necessity in mind.

The manifold metamorphoses of form in moving water that occur in conformity with cosmic events show too how strong the interplay between the earth and the planetary world can be, an interplay that manifests itself in visible flow-forms, and how actively water mediates between the two realms. Wholly new approaches to an understanding of water as the life-element can be opened up on the basis of these facts.[7]

Studying the Behavior of Water

Wolfram Schwenk

In recent decades, a decline in the quality of drinking water immediately perceptible to the senses has been noted in a great number of densely populated areas of industrial countries, and this despite the fact that the water supplies in question were in every case unobjectionable from a hygienic standpoint and were therefore adjudged acceptable. This observation raises the question what scientific methods could be added to the analytical for purposes of judging and differentiating between such waters from the qualitative aspect reflected in human sense impressions.

Hygienic-analytical criteria and testing methods investigate the components contained in water. But what we perceive of a water sample with our senses and how we feel about it is not just a matter of its components, but of a total impression that the water composed of them conveys. It is the way water behaves that calls forth the impression of its pristine springlike quality or of its deadness.

Government officials and experts have thus far neglected the entirely legitimate question about water's behavior as a standard upon which to base qualitative judgments.

Persons whose responsibility it is to judge a building material inquire as a matter of course into its compo-

Reprinted from *Der Aufbau*, No. 3–5, 1979.

nents and their composition. But on the score of quality, they assign every bit as much importance to testing its technological and physical aspects, its behavior. Vital criteria such as the material's capacity to insulate against cold, noise, and moisture, its brittleness or elasticity, its reversible or irreversible malleability when subjected to heavy loads or to heat, are decisive, as expressions of the behavior of the material, in judging its quality—each one a pebble contributing to the whole mosaic.

Certain building materials are known to behave similarly or identically though consisting of very dissimilar components, while there are others that, though their composition is identical from an analytical standpoint, behave very differently if their component parts are differently proportioned.

What is the situation in regard to the behavior of water, the most universal building material and life resource of organisms? Are there common aspects and differences in the behavior of potable waters?

The most elementary experience we can have of water and the one we most clearly connect with its behavior is its inner mobility. Water, like all fluids, presents an integrative expression of its behavior in its movements, and in the patterns these movements momentarily call into being.

As we learn from hydrodynamics, the interplay between endogenous and exogenous factors of a system influences the movement taking place in it in a characteristic way. Characteristic flow-forms appear as the result of the interaction of all the factors involved. They show, in addition to the details that can be measured in figures, a total graphic picture of the situation with all its particularities. Especially in the labile border area between the laminary and the turbulent flow, which can be exactly adjusted by a corresponding choice of exogenous factors of the system, the most minute changes in a fluid's attributes register in a change in the flow pattern, whether this moves in the direction of an undifferentiated laminary or of a more richly differentiated turbulent flow.

Here is revealed a connection in the case of which the motion of the fluid with its flow-forms becomes a perceptible pictorial expression of its quality, a quality manifest as movement.

The drop-picture method developed by Theodor Schwenk is a procedure for the comparative investigation of the quality of different water samples based upon these connections.[1]

In this procedure a thin layer of the water to be examined, which is motionless in a level glass bowl, is brought into motion by a certain number of drops of distilled water falling into it. The flow-forms thus created in the water sample (a homogenous solution of the sample with the addition of glycerin as a neutral, inert streak-former) are rendered visible by means of schlierenoptical devices (with the help of the Töpler schlierenoptical apparatus) and photographed.

In testing water, it is expedient to undertake the calibration of the drop-form apparatus with distilled water available in consistent quality. The flow-form patterns remain almost unchanged for several seconds until the next drop falls in. The succeeding patterns go through a characteristic evolution. A rosette, the outer leaf-form elements of which are composed of single vortices, is only delicately suggested after the addition of the first drop, and develops from drop to drop in a gradual expansion, multiplication and differentiation of these vortices. After the twentieth drop, this development slackens, and new, dendrite-like structures appear, which determine the pattern from about the thirtieth drop on.

A further addition of drops leads by way of various transitional stages to phenomena of stepwise dissolution, accompanied by the formation of an undifferentiated disk.

This normal type of an evolutionary sequence in the drop-pictures can easily be reproduced if the extremely subtle way of proceeding that is vital to the method is maintained. The sequence undergoes a great variety of modifications when a number of other water specimens are similarly brought into motion under identical experimental conditions. (See pictures II–IV and XIV. All demonstrate the developmental stage of the water samples at the twentieth drop.)

The drop-pictures of natural, untainted ground water, of ground water wells, and of brooks in the pristine spring region all belong, practically speaking, to the normal type and appear as reproducible modifications of that type. (See picture XIV and III) Among them are water samples of the most varied hydrogeological origin and hydrochemical make-up. Mineral water specimens are the only ones that deviate significantly, each in its own specific way, from this basic type. Water samples that transport a load of foreign matter such as sewage or effluents produce less thoroughly detailed and structured flow-forms, and culminate in the monotonously disk-shaped type which puts in an appearance already at a low drop-count. (Picture IV) They are examples of a shift in the balance of factors brought about by the inclusion of disturbing substances that shift the flow toward the laminary side. They move sluggishly.

Current measurement technology allows us to trace how the three main physical parameters: surface tension, viscosity, and density contribute both individually and in their interaction to this or that type of structure.[2] In the case of a series of dilutions, even when the accuracy of the figures of these factors allows for no further distinction between the specimens, changes in the drop-picture from specimen to specimen can still be discerned in the development of the flow-forms up to dilutions in several series of powers of ten.

Thus far our experience indicates that no matter-specific flow-forms show up in tests of drinking water by the drop-picture method. This method is therefore not a substitute for analytical methods. It simply supplements them. It proves its value as a testing procedure in border areas, for example, where it can take on the function of an indicator of summarizing parameters. Picture II is the drop-picture of the drinking water of a large West German city, water sensed by its inhabitants to be poor in quality and recently discovered by hygienists, in a further refinement of their analytical procedures, to contain undesirable substances. Purely on the basis of its greatly weakened capacity, compared with that of spring water, to structure the flow-form, the drop-picture of this specimen reveals the fact that disturbing elements are present, and this at first glance, not after extension of the testing program.

The differentiating capacity of the drop-picture method extends from examples such as the above to the determination of subtle differences that exist, for example, between drinking water that has remained stagnant for several hours in a water main and the same water a few minutes after flowing through it.

Drop-pictures also make it possible to distinguish clearly between big-city water samples collected on workdays and on holidays. Natural ground water suffices for holiday drinking water, whereas on workdays there is an additional call for water for industry, business and public institutions. This makes it necessary to supplement the supply of ground water by mixing it in interconnected mains with water from filter-stations on the banks of effluent-laden rivers.

Examples of this kind bring home the fact that the drop-picture method can determine the characteristics of a water specimen only at the moment of taking the sample, and that here too, as in the case of the analytical method, samples have to be tested over a fairly long period of time to make a generally valid judgment about the water of a particular locality.

Experience gathered in many such drinking water flow-motion studies demonstrates that the subtle qualities that people note in their synthesizing perception express themselves in reproducible flow-forms. The drop-picture method can therefore contribute to an objective determining of subtle qualitative gradations in judging various hygienically acceptable water specimens where analytically derived data alone still provide no basis for judging their overall condition and special attributes. Thus, this method sees itself charged with a task presented by the prevailing demand for good drinking water.

According to the general regulations embodied in the West German *DIN 2000 (DIN* = German Industrial Standard), the requirement for acceptable drinking water from central supply sources is "to come up to the standard of acceptable ground water derived at a sufficient depth from sufficiently filtered veins of naturally circulating water that has not been exposed to any contamination whatsoever."[3] But hygienic regulations actually restrict themselves consciously to prohibiting what is either unallowable in drinking water or allowed only in certain specified concentrations. The drop-picture method adds the necessary further picturing of a water specimen's behavior as a result of everything it actually contains.[4] Without it, the required comparison would be limited to necessary concomitant circumstances, ignoring the question of the character of the water itself.

The suitability of using water's dynamic reaction to make this comparison is clearly evidenced in the fact that chemically very different ground waters that in their entirety form the standard of acceptability under *DIN 2000* regulations reveal a common basic type in their drop-picture motion-forms. On the basis of their behavior, they provide an objectively demonstrable standard of comparison.

The fundamental type can therefore serve as the basis of comparison for acceptable, spring-fresh water. Comparative studies of the drinking water of a great variety of localities in West Germany and Northern Switzerland undertaken for this purpose by the Institute for Flow Sciences showed a consistently good spring water quality in water delivered by small-town mains from their own ground water or spring-fed wells when the water had been allowed to run for a while. Only after longer periods of stagnating in the pipe system was a relatively lesser quality observed.

In larger towns and especially in big cities supplied by filtered surface water, deficiencies in quality were found repeatedly, even in running tap water. In localities of this kind, the quality of tap water was always surpassed by that of water from permanently flowing springs. This fact should be a sufficient indication of the desirability of supplying the public with artesian well water wherever possible, for the many people who reject their town tap water for nutritive purposes. The habit, widespread among officials of West German water departments, of declaring, as a purely precautionary measure, that such well water is unfit for drinking (for the sole purpose of avoiding the cost of constant hygienic supervision) should not masquerade as the right approach.

The drop-picture method was developed for the specific purpose of testing drinking water reaction, but it can also be used to examine other types of water and fluid solutions in general.

In testing surface waters, it is impressive to see how the wealth of patterns observed in drop-picture studies of certain river sections parallels the variety and organizational levels of the sequence of life-communities in the corresponding sections.[5] Drop-pictures also make it possible to discern how water responds in the stepwise processing of surface waters into drinking water.

Medicinal and mineral waters evidence the greatest differences. Studies of sea water, of liquids such as beverages, juices, pharmaceutical preparations, body fluids and the like are possible, but a comparative standard must first be established to support any conclusions.

The interpretation of drop-pictures is arrived at purely empirically, on the basis of experience. It does not proceed deductively, but neither does it deductively exclude new findings.

Water-motion study is indebted to this empirical procedure for new insights into relationships in nature that point to a close correspondence between the earth and the planetary world. The apparently wide distribution of experimental findings in cases of turbulent flow is closely connected with events in the stars; taking this fact into account, this wide distribution can be considerably narrowed down by carefully calculating the right moment for testing.[6] For efficient use of the means of research, it is necessary that scientists working with the drop-picture method take such relationships into account and make use of them. A wide range of as yet unpublished findings has been amassed; their consequences for the experimental work will of course be communicated to trainees working to gain facility with the drop-picture method.

To sum up, it may have become obvious that a wide perspective has opened into new and unusual territory with the investigation of motion patterns in water as practiced for example in the drop-picture method, and that a great deal is still to be learned about it.

The Experiment Station of the city of Vienna looks back upon a century old tradition of boldly exploring new areas of science and immediately turning its findings to practical account. As a result of experience derived from its consistently future-oriented tradition, it has joined forces with us in this youthful and unconventional research and is working at it in an admirably reliable way. Its efforts rank among the most outstanding in this research field. Heartfelt congratulations and best wishes for the future from the Institute in which this line of research, based on Rudolf Steiner's spiritual science, was launched!

Water as a Nutrient

Wolfram Schwenk

You go to a wall panel in your home, turn a faucet, and—lo and behold!—a nutrient comes pouring out of the end of a pipe: drinking water, the most important nutrient there is. You need not feel the least concern about it, for in civilized countries you may be certain that it is potable, unexceptionable, that is, from a hygienic standpoint.

What other foodstuff flows into our homes, free of charge, in apparently unlimited amounts? And since we come by water so easily we forget how precious it is. We put it to use quite as a matter of course for other purposes besides ingestion, for a great many other purposes indeed, with which we are all familiar. Nor do we use it sparingly. Quite the opposite: we use it to excess. More than fifty times as much water as we need for drinking, that is, for nutritional purposes, is lavished on processes and problems on which a nutrient should not be wasted, especially not the most important, indispensable nutrient we have. But society and industry want this misspent 98% of our water to be not only just as unexceptionable from a hygienic standpoint but also to measure up to the highest drinking water standards.

This has consequences. Since nature does not provide

This article originally appeared in the German periodical *Die Drei*, No. 7/8, July/August 1982.

water in the amounts currently demanded,she is being plundered. Ground water supplies are being exhausted, and either not replaced at all or only in the form of polluted, poisoned surface water, which is not good enough for our use.

The problem of water contaminated by chemical effluents from farms and factories that was thought just a few years ago to be one of polluted surface waters is only now confronting us in ground water even at great depths and in remote regions, and in evermore threatening degree. Among these pollutants are the residues of highly toxic solvents, poisonous chemicals, and heavy metals from industrial plants, which either get directly into water from dumps, sewage, street cleaning and the atmosphere, or else into the soil. They are dissolved there and are carried down into the ground water by acid rain. The same thing happens as a result of the spreading of chemicals over farmland in the form of mineral fertilizers, exceedingly toxic pesticides and the like, not to mention long-lived hormones such as estrogen from animal feeds, found not only in veal but for years now in ground water as well.[1] A little over a decade ago, ground water was laden with nitrates from farm use in just a few especially famous wine-producing regions, but now these residues are commonly found in all intensively farmed areas. The techniques applied in treating drinking water have thus far failed to come up with any practicable procedures for the selective removal of nitrates from water supplies. The tolerance level was accordingly raised for the time being. But increases in the nitrate content of drinking water are a potential danger to human health, and for infants they are a threat to life itself.

Organic-biological farmers as well as those who apply mineral fertilizers are endangering the ground water with their farming methods.[2] Only biodynamic agriculture with its composts makes use of fertilizers so stabilized as to offer no threat to ground water. From the standpoint of water departments this agricultural method is ideal and should, if only for that reason, be supported publicly and disseminated without delay.

After decades of successful efforts in the treatment of biodegradable domestic sewage, the threat to water supplies comes not so much from that source as it does from the countless invisible water-soluble chemical compounds contained in effluents and exhaust gases produced by industry and its customers.[3] The only reason why these substances are not officially designated toxins is because it has been impossible for the testing authorities to keep pace with the constantly increasing rate at which they are being produced.

Pipelines and ships yearly dump some 100 million tons of industrial wastes near the coasts and into several areas of the high seas, and this is legally permitted. The sea is contaminated thereby to such an extent that it is common to find fish severely crippled or with skin erosions—if, indeed, they are found at all.[4] Can we under these circumstances expect the sea to serve as the food-source of the future?

Let us not forget that industry produces only what we will buy. The environment is made to bear heavy burdens by our senseless squandering of all kinds of paper and packaging materials. This does not only occur at the dump. Approximately 300 liters of fresh water are needed in manufacturing every single kilogram of paper in the medium quality range and are transformed in the process into troublesome waste-water. More than 400 liters of the best fresh water are required per kilogram of finished product in making glossy paper of the short-lived kind favored for catalogs, advertising, appeals and the like, and become problematical effluents. Recycled paper, on the other hand, requires less than 2 liters of fresh water for every kilogram produced.[5] All recycling of paper greatly relieves the burden on our water supplies.

It is no longer possible to regard problems of the overburdening of soil, air and water with health-undermining and even life-threatening chemicals—for which man must answer—as isolated or locally limited cases, or to solve them in isolation. Their synergistic effect has created a global catastrophe. As a result of air pollution over extended areas, rainwater in Switzerland, for example, contains on a national average as great a content of dissolved filth and poisonous substances as the sewage-laden Rhine does at the point where it leaves the Swiss city Basel.[6] Even more drastic than the dying out of pines here at home is the dying of numerous bodies of water in Scandinavia as a result of countrywide sulfuric acid contamination, originating in the heavily polluted air of central and western Europe. Instead of undergoing cleansing at the place of its origin, this contaminated air is dispersed through the atmosphere by tall chimneys in such a way that precipitation carries the poisons over whole continents.

Where can pure water be found to dilute rain of this kind? What soils are still so undamaged and so low in salt content that they are able to filter such polluted water successfully? It will not be long before unspoiled water is sought for in vain. The worldwide water catastrophe of coming decades has already started.[7] And water catastrophes are life catastrophes and cultural catastrophes.

Putting water to use in the service of our highly civilized culture with all its amenities has contributed greatly to the solution of technological problems that have surfaced, but it has also led to the kind of thinking that understands water only as a raw material to be used in meeting a tremendous variety of competing industrial needs. It is this attitude of mind that is causing water to become estranged from the life-element. With no other nutrient do we behave as thoughtlessly and inconsiderately as we do with water, the most important nutrient there is. People in our times have unlearned and neglected to develop and value insight into the nature of this life-element. Worldwide pollution of wells is the result. Water is just the place where it becomes evident that human beings can only destroy the bases of life with their materialistic, analytical orientation, for they lack the tools of knowledge required for their nurturance; their traditions have become hollow, and the old instincts no longer give them any support.[8] The devastated environment all around them already reflects the devastated nature of their thinking.[9]

If, against this background, we examine the motives of those responsible for the ubiquitous large- and small-scale contamination of water, we become aware on the one hand of laziness, thoughtlessness, and conceit, and on the other, of solid economic arguments—in both cases, unintentionally or otherwise, a lack of awareness and of responsibility for the total organism of which man and nature both form a part. The branches of industry that take the trouble to eradicate the symptoms are those with exceptional growth prospects, where political expediency is involved. But superficial efforts to repair the damage merely shift the problem. The worldwide pollution of the environment is symptomatic of unsolved questions affecting the whole human race. These questions call for a new attitude, for solutions springing from a reorientation based upon a new view of mankind and the universe, of the way the whole human community conducts its common life; they call for the threefolding of the social organism,[10] for true stewardship, the fruit of insight into the life-element,[11] in caring for the earth-organism, and for practical work with Rudolf Steiner's spiritual science.

But drinking water pours nevertheless in potable condition out of pipes in our homes. If you were to express any concern about it, officials of your water-processing plant and the health authorities would tell you that it has been tested and approved. What is there to worry about? Everything is taken care of in the best possible way; no objection can be made on hygienic grounds.[12] You will be told that your impression that the drinking water in one place is refreshing and alive, but dead and insipid in another is totally unfounded and subjective, not susceptible to scientific measurement. It is still quite possible, using expensive technological methods, which include biological procedures, to turn a good many tainted water supplies into potable water; the experts perform near miracles. Only by continually adapting and developing their technology further are they able to remove ever new contaminants from water supplies as they process them for potability to the extent officially required—a race without any ending.[13] The regulations, which owe their existence to unfortunate experience, are by their very nature retrospective and necessarily full of holes. Regulating laws, applied under strict supervision, cover requirements for drinking water that may seem unobjectionable from the hygienic standpoint.[14] Potable water is not allowed to contain any living microorganisms originating in human or animal excrement. Poisonous or esthetically repellent substances either may not occur at all in drinking water or else only in specific concentrations regarded as safe. The safety or tolerance level is established by a series of bacterial tests, by five physical as well as by 12–30 chemical determinations, made according to standardized testing procedures on a standard selection. Anything that is not caught in the strainer of these standards passes through and remains unidentified; an unrecognized and varying but considerable number of finely diluted poisons scarcely obtainable without a doctor's prescription if one wished to consume them in like amounts in a pharmacist's medicine bottles.[15] A further requirement made in these technical regulations is that drinking water be cool and appetizing: "A water is appetizing when its external aspects together with its physical, chemical, microbiological, and biological properties show no sign of contamination, and provided that there is nothing repellent about the method whereby it is obtained."[16] (*DIN* 2000, Section 3.3.1.)

From a hygienic standpoint the purity of water is a minimal requirement, which it is the duty of the experts on water to meet. But it is the product of an exclusively negative concept of quality, and it covers only those properties that drinking water must not possess under any circumstances. No one questions such a regulation; on the contrary; it is not nearly strict enough. But there is another aspect too in which it is inadequate: the absence of undesirable components by no means confers the desirable degree of excellence on a nutrient, though many branches of the food industry assert the contrary. Drinking water, the most important of all nutrients, has not been ranked among foods in the matter-of-course way in which all other foods are so characterized.[17] The life-giving, life-serving properties of fresh water need to be investigated and described, and these properties must become the yardstick in judging quality.

There has been no lack of effort in this direction. The guidelines for regulating public water supplies embodied in *DIN* 2000 require that the standard of quality measurement for wholesome drinking water be based on the model of unspoiled ground water. (6, Sec. 2.3.). But what characterizes this model? As a rule, unspoiled ground water is water sensed by us to be not only drinkable, but "good" and "alive." How can this impression be characterized? One thing to keep in mind is that the living quality of water is not the product of opinion or of a way of looking at things; it is an immediate experience of our most vital nutrient, an experience of which we are most aware when the drinking water coming out of our faucets has lost its spring-like freshness and life, and we have to get along without them; when, in other words, we feel, every day anew, a distaste for what is coming out of the faucet.

What is the expert's approach to the requirement? He reduces the question, in a way typical of the current scientific habit of thinking, to testing what substances are present in "good" and "bad" drinking water. Top-ranking specialists in water treatment have performed tests and made comparisons of this kind, and have found a fairly similar distribution of mineral components in both.[18] Their conclusion is that the sensations "good" and "bad" are not susceptible to scientific proof, and, further, that it is not possible to characterize water quality positively. This generalized judgment resulting from basing the question one-sidedly on components demonstrates the grave inconsequence of this way of thinking, expressed too in such unrealistic concepts of quality as that of the description "appetizing" referred to above. Qualitative judgments like good/bad, refreshing/insipid, living/dead are the product of a synthesis of the great variety of ways in which the senses perceive the properties of water.[19] These are not summed up in answers to the question as to what is present and in what amounts. We must go further and look into the interplay of the component elements, seeking answers as to how they behave, for it is not so much these components that people experience and judge as it is the totality resulting from their interplay. And every such composition is qualitatively more than the quantitative sum of the components separately analyzed. The analytical sifting and enumerating of the components may contribute to the question how water behaves, but the analytical method can never produce a satisfactory answer. Scientists may object that the point here, as in the case of living organisms, is simply that many different kinds of things are involved in a highly complex interrelationship, and that the problem this presents for the analytical method is only a matter of degree, not of principle. But complexity is not the only decisive aspect, as the difference between a living organism and its dead remains demonstrates, since, viewed from a purely material standpoint, both are equally complex in structure. Analysis is not the only suitable tool in answering the question of the *how* in this case; it is sufficient only for the *what*. It is indispensable in guaranteeing the absence of certain undesirable components by testing, but it is unable to disclose the properties of a given water in the interplay of all its components. For this, supplementary synthesizing ways of testing and characterizing are essential.

When we are dealing with concepts of quality, based upon a total impression of how the components work together, the presence or absence of certain ingredients is not the only thing to be considered; the thing that matters is what is happening and what is active in the components present. All such impressions of quality—

expressed in music as consonant or dissonant, in nutrients as good or bad, delicious, refreshing, insipid, or unappetizing—call forth a feeling response in the person affected, as the adjectives suggest; they indicate a sympathetic or antipathetic response. As Goethe pointed out, they affect the ethical sense as well as the physical.

If we want to transform our subjective impressions into objective certainty and to express them in clear concepts, we have to practice activating our sense perceptions and, in disciplined observation of our own perceptive activity, develop the capacity called by Goethe "the power of perceptive judgment." The consistent application of Goethe's scientific approach, whereby the physical-ethical effects of things can be observed and judged just as exactly and reliably as so-called objective data devoid of ethical values, is based among other aspects on the fact that differences between things are studied by contrasting them and comparing their common elements rather than by splitting them up into their separate components. When this synthesizing Goethean mode of studying things is accorded its equally justified place alongside the dissecting, analytical method and used in combination with it, proper answers can be found to our questions concerning the living quality of water.

Let us now describe water with concepts derived from everyday life. We find it possesses no form of its own, no odor or taste, no color, and no tone. The concepts that we form of the material world on the basis of our five senses either do not apply to water at all or do so only in a negative sense. And it is just this lack of concepts applicable to the uniqueness of water that makes an approach to the essential being of this element so difficult. The method whereby that being can be suitably investigated therefore has very little in common with the material approach usually practiced, and we are therefore challenged to develop new perceptive capacities. This holds as true in the case of a cultivation of personal perceptive activity and capacity as it does in the developing and cultivating of new methods of natural scientific research. Rudolf Steiner spoke of both as urgently needed, and he gave fruitful indications for both ways of proceeding.[20]

We will devote ourselves here to pursuing the second procedure, the path upon which the various image-creating methods came into being in response to Steiner's suggestions. These include Ehrenfried Pfeiffer's sensitive crystallization method, Lily Kolisko's capillary dynamolysis, and the drop-picture method developed by Theodor Schwenk. In all three cases, an opportunity is created by means of experimental set-ups and procedures for a shaping process to take place in the solutions under study. The resultant patterns are compared, described, and judged, applying Goethe's "power of perceptive judgment." These methods meet the requirement that exact, comprehensible natural scientific phenomena be brought together with the human being who applies them; without human inclusion they are worthless.

In a lecture given by Rudolf Steiner on April 1, 1921 in Dornach, Rudolf Steiner expressed himself very strongly on the significance and necessity of establishing this type of connection between the researcher and the material under study.[21] And F. Hofmann, who fails in other respects to do justice to Goethe's scientific work, says in an article on Goethe written for the 150th anniversary of his death, "Just today, at a time when the discovery of isolated natural scientific facts such as atom splitting or the genetic code evoke threatening visions of the future, we should think back to and ponder Goethe's insight that it is 'the worst calamity in modern physics that experimentation has become, as it were, divorced from man.' In spite of his scientific mistakes and blunders, Goethe must be given credit for having attempted a synthesis between man and science."[22]

All three image-creating methods have their own particular focal point of sensitivity and qualification for certain areas of study. In the investigation of water it is the drop-picture method specially developed for this purpose by Theodor Schwenk.[23] It works with the element of movement in water and brings to visibility the flow patterns created as water moves. The specimen of water under study rests in a flat round horizontally-positioned glass dish, as a thin layer approximately 1mm deep, and is set in motion by drops of distilled water falling into it at regular five-second intervals from a height of 10 cm. The resulting movement is rendered visible by a schlierenoptical technique. This consists of Töpler's schlierenoptical device, casting a beam of light, and uses a standardized homogeneous dissolved admixture of glycerin in the water sample. Every drop brings into being a new flow pattern of vortices, which spread out horizontally in all directions, creating a rosette form. As the number of drops increases, the patterns are further differentiated, and a development takes place involving a gradual change in the type of pattern. These flow-figures are metamorphoses of vortex-rings.[24] With the addition of further drops, linear structures appear in largely radial arrangement. Unlike the vortices, these are not completely renewed; they remain as before for awhile, and then undergo only partial change and extension with the addition of still further drops. Sooner or later the earlier rosette-shaped vortex-forming type of figure is replaced to a considerable extent by the second type. In the transitional stages a progressively creative type of pattern-forming wrestles with the conservative-additive type, each type manifesting itself more or less strongly depending on the water sample studied.

If the experimental conditions are rigorously standardized and carefully managed, the flow patterns can be reproduced. In that case, the developmental course of the series of patterns results in modified pattern sequences and pattern-formation specific to the water sample under study.[25] This makes it possible, by means of comparison, to delineate certain qualitative aspects of water.[26] From the rheological standpoint these interrelationships are supported to some extent by the Reynolds formula for determining the similarity of flow patterns. Ascertainable cause and effect relationships are described in the technological literature for rheological and physiochemical parameters.[27] What we are concerned with here is the fact that a reproducible expression of the interaction of all the components and factors in a given water sample comes into manifestation in the behavior of moving water, reflected in the drop-picture, and that this makes possible the desired pictorial representation of the total behavior of the water sample.

Now how is the living condition that is our immediate experience when we taste really pristine water manifested? In the case of good unspoiled ground water, of water from springs fed from such sources or from the mountain brooks that such untrammeled springs give rise to, the vortex-rosettes appearing in the drop-pictures consist, up to the twentieth drop, and further, of multiform vortices in rhythmical, coordinated arrangement. A variety of well-developed single forms appears, which, despite their great individual differences, combine to make a harmonious whole—a diversity that does not split up into proliferating details but develops within an integrated framework.

The flow patterns of such waters do not, then, immediately follow the physical laws of thermodynamics according to which one would expect them to move toward the most likely outcome: that of maximum chaos. Instead, they overcome that tendency temporarily, and flow in movements in which new, intensely organized rhythmical shapes come into being. Pictures of this sort exhibit a living organization. Their effect, from the esthetic standpoint, is one of life and beauty. Comparing them on the basis of a chemical analysis of the water samples under study, one finds no correlation with the material components. But the picture type of the coordinated rosette resulting from the manifold differentiations and structures of the individual vortices nevertheless prevails in the case of all water samples with a subjective correspondence to the standard for acceptable drinking water imposed by *DIN* 2000 regulations, even allowing for the individual characteristics of the various samples. This type of picture does not occur in the case of samples taken from tainted ground water or from surface waters that, after being processed and rendered hygienically acceptable in waterworks, are still sensed as "bad."

This demonstrates the fact that the drop-picture method is capable of delineating the prototype of acceptable water established by *DIN* 2000 regulations, and that objective comparisons of various drinking waters with that standard can be made with the aid of this method. It must be emphasized, however, that flow patterns cannot be substituted for chemical analysis and bacterial testing, for they do not provide information about the *what*. Hygienic and analytical testing is still absolutely necessary. But once it has been established on the basis of the *what*, of content, that a given water is acceptable for drinking, the pattern-creating drop-picture method can supply additional information as to whether and to what extent that water corresponds, on the basis of the *how* of its behavior and life-qualities, to spring-fresh, untainted ground water. Certain types of patterns are connected with the increasing staleness of drinking waters, and various transitional forms can also appear between these. Examples are shown following page 191. In the case of all these stages, a centrally oriented symmetry can still be discerned, along with a stepwise loss of differentiating capacity. In other cases, the most varied form elements can proliferate singly, and then show up in exaggerated formations in various parts of the patterns, having lost all coordination.

According to the conceptions of modern natural science, these phenomena can be described as an incapacity, gradually growing more evident, for waters of this kind so to respond with motion to the kinetic energy introduced by the fall of the entering drops that variously structured and coordinated "information-rich" forms come into being. At this stage of current abstraction one can therefore say that *living water*, flowing in motion where new, highly differentiated coordinated structural arrangements originate, overcomes the tendency to strive for the most probable condition of chaos; it is, in other words, information-rich, negentropic. *Bad water* quickly tends toward the most probable condition of maximum disorder, that is to say, it behaves in an information-poor, entropy-increasing way.

These facts indeed entitle us to apply the concepts

"dead" and "alive" to the waters described in accordance with the abstract view of life held by modern theoretical biology, but we can retain and apply them equally well in accordance with the very newest scientific outlook—and not just based on subjective impressions, but on the basis of empirical experimental findings. This contradicts the assertion of water experts that the impression of a living quality of water is not scientifically valid and cannot be demonstrated. From now on, such statements must be regarded as pretexts.

Let us now proceed, within the framework of the overarching theme of this article, to discuss some further basic aspects rather than cite single instances and aspects of various drop-picture modifications met with in particulars of drinking water processing and in regard to special functional peculiarities of certain waters.[28]

Who or what creates the wealth of differentiated, highly organized motion patterns that convey a picture of aliveness to us? As I have already reported in the case of specimens of acceptable waters, we find no connection between these patterns and the physical substances they contain in solution. In the case of less acceptable waters we have now come to know that disturbing organic-chemical substances can bring about a weakening of formative capacity, particularly as a result of lessened surface tension. But the demonstrable active presence of the components dissolved in the water extends only to disturbances of motion, and weakenings of form not related to any specific component. They are not, on the other hand, related to the organizing, constructive building up of form; substances in solution hinder rather than further this process. Its causes are not to be found in the crudely material, analytically determinable *what*.

We may say that the following observations of forms that appear momentarily in flowing water hold true in general, not just in the experimental special case of the drop-picture method. On the basis of the manifold phenomena of richly structured picture-forms that come into being in flowing water as a result of movement rather than being determined by material components, we are entitled to speak with full scientific justification of the activity of organizing formative forces, forces that work into matter rather than proceed out of it. The support for this statement and further characterization of the forces referred to are presented and treated in detail in a separate article.[29] They are no less firmly substantiated than the assumptions of theoretical physics on the "self-organization of matter," which is supposed to be primarily "responsible" for all macroscopic formations in the molecular realm, though the scientists in question certainly make no specific pronouncements on the score of this (atomic) "self."[30]

Substances that act as the vehicle, the carrying medium, must of course be present as the precondition for the activity of the formative forces in the physical-material realm, both in the form of water or dissolved in it. They must be conceived as the enabling condition, not as the cause; they simply provide the opportunity. The vehicle water is activated by movement, which brings internal plane-like structures into being in a tangential sheath-building arrangement.[31] Under these conditions formative forces in water can act to introduce form, bringing movement patterns there into transitory ordered, organized form, a process ever freshly and rhythmically repeated.

Theodor Schwenk has shown how the movements that occur in the development of organs and the types of organic forms thus created are reflected in flowing water in the patterns that come into being there, so that these patterns may be described as organ-related flow-forms.[32] This aspect provides quite special justification for speaking of the activity in flowing water, too, of formative forces that carry on their creating in the living realm. These facts bear witness to the life-serving, life-renewing activities and effects of fresh water.

Now let us return to a discussion of our studies of water specimens. We can learn from them—on the basis of the picture sketched here—to read the differentiated activities of the formative forces at work in the various waters studied. In the living realms of nature, this differentiated weaving of forces is directly reflected in the single types, species and organs of living creatures. In the case of the fluid mineral substance water, the pattern-creating method is required to provide a corresponding opportunity there for a picture of this weaving of forces to appear in physical manifestation. However, we recall here Rudolf Steiner's indications that what is at work here is not some anonymous force or other; individual beings of various hierarchies and elemental kingdoms are involved and active in them.[33]

These activities that can be read from water patterns are not only found in the various water specimens as an expression of their characteristics. They can manifest themselves differently in one and the same water specimen, seen from the standpoint of its crudely material make-up. In such cases, these characteristics reflect qualitative aspects that change with the passage of time. The flow-forms in a sample of living water constantly modify their normal reaction in accordance with the position and movement of the planets in a more or less intensive process of strengthened or weakened formative power. They tell us in a physically perceptible way about the cosmic origin of the formative forces, confirming

Steiner's indications that fluid substances such as water "stand under the influence of the whole planetary system," and that the forces of the various planets "take effect from the part of the heavens where these planets stand."[34]

These connections can be demonstrated by careful choice in the matter of timing and by careful experimentation, always distinguishing them from statistical distribution patterns. And they can be deliberately kept out of manifestation by finding times at which the normal behavior of the water can be tested.

Now what does all this have to do with the nutrient properties of water?

Water's role as a vital nutrient can actually be estimated only on the basis of these findings. We know from experimental, empirically obtained findings that living waters are open and receptive to influences raying in from the cosmos; their flow-forms reflect the changes that the passage of time brings about in cosmic moods and qualities. They are connected with the source of the life-bestowing formative forces, and mediate these to living creatures.

Bad waters, deficient in formative capacity, are as though deaf, apathetically insensitive to the attributes of the cosmically ordered time-stream; they are incapable of change. For this reason they are unable to channel formative forces from their source to living organisms.

Does such mediation to living beings actually take place? This is a fact that has been demonstrated by Theodor Schwenk in the growth of seedlings in water specimens subjected to various kinds of exposures. The many-faceted findings produced by his research have shown in addition that living waters respond not only selectively[35]—as certain material solutions do to certain planetary events—but respond universally to the most varied planetary influences and can pass them on.[36] Extensive research as to the nature of these effects is presently being conducted at the Institut für Strömungswissenschaften, Herrischried, West Germany.

Acceptable waters are in this sense transparent, as unacceptable waters are not. When, misusing water, we ruin it and render it insensitive to cosmic influences, we close not only ourselves, but the whole world with all its life-realms, to the cosmic life-source. Then devastation is apparent, in and around us.

It lies in our power to decide whether water can mediate between these two worlds, whether it can be a true or a specious "conveyor of life." This decision begins at the water faucet as well as in our own choice of what paper to use, not in a curing of symptoms. Insight is not enough; if we are to act freely and responsibly rather than follow a prescription laid down for us, we will need perceptive judgment. This is the challenge to every one of us, not just to the professionals. The capacity for perceptive judgment is founded upon the ability to perceive and to experience. For how otherwise than through the gate of warm inner experiencing is it possible to achieve a moral relationship to the environment?

Notes

What Is "Living Water"?

[1]Johannes Kepler (1571–1630), German astronomer who discovered Kepler's laws of planetary motion. Postulated ray theory of light to explain vision.

Guenther Wachsmuth, *Etheric Formative Forces in Cosmos, Earth and Man: The Path of Investigation into the World of the Living*, vol. I, 2nd ed., (New York: Anthroposophic Press, 1932).

Walther Cloos, *The Living Earth*, (London: Lanthorn Press, 1977).

[2]August Schmauss, "Biologische Gedanken in der Meteorologie," *Forschungen und Fortschritte*, 21 (1945), no. 1–6.

[3]Paul Raethjen, *Dynamics of Cyclones*, (Leipzig, 1953).

[4]See Theodor Schwenk, *The Basis of Potentization Research*, (Spring Valley, N.Y.: Mercury Press, 1988); *Bewegungsformen des Wassers: Nachweis feiner Qualitätsunterschiede mit der Tropfenbildmethode*, (Stuttgart: Verlag Freies Geistesleben, 1967), not yet translated; and *Sensitive Chaos*, (London: Rudolf Steiner Press, 1976).

Water Consciousness

[1]See Theodor Schwenk, "What Is 'Living Water'," pp. 1–13 of this volume.

[2]Lord Francis Bacon, First Baron Verulam and Viscount St. Albans (1561–1626), English philosopher and author. In his work *Novum Organum* (1620), he presented a systematic analysis of knowledge, intended to replace the deductive logic of Aristotle with the inductive method in interpreting nature.

[3]Johannes Kepler (1571–1630), German astronomer who

discovered Kepler's laws of planetary motion. Postulated ray theory of light to explain vision.

The Spirit in Water and in Man

[1]Plato (c.28–348 or 347B.C.), Greek philosopher, student of Socrates and teacher of Aristotle. One of the founders of Western philosophy and culture.

Aristotle (384–322B.C.), Greek philosopher and student of Plato. Later tutored Alexander the Great. Author of several books on philosophy, politics, and culture.

Boethius (c.480–524), Roman philosopher who translated several of Aristotle's works and wrote commentaries on them. Author of various treatises on music, logic, and theology.

Augustine (354–430), known as St. Augustine of Hippo. Early Christian church father and philosopher. Converted to Christianity after passing through a spiritual crisis. Author of influential writings and later bishop of Hippo.

[2]Alanus ab Insulis, Latin name of Alain de Lille (c.1128–1202), French philosopher, theologian, and poet.

Albertus Magnus (c.1200–1280), also called Albert the Great and Albert of Cologne, German scholastic theologian, philosopher, and scientist. Entered Dominican order and taught in Paris and Cologne, where Thomas Aquinas was one of his students. Wrote many treatises on scholarly and scientific subjects, attempted to unite theology and Aristotelianism, and succeeded in establishing a place for natural science in system of Christian studies.

Thomas Aquinas (1225–1274), Italian religious and philosopher, entered Dominican order. Student of Albertus Magnus and outstanding figure of scholastic philosophy, integrated scientific rationalism and naturalism of Aristotle with Christian revelation and faith.

Averroës (1126–1198), Islamic philosopher, known in Europe under this name. He was a principal interpreter of Aristotle's writings, influenced later Jewish and Christian

Aristotle's writings, influenced later Jewish and Christian writers, and worked to reconcile Islamic and Greek thought.

Anselm of Canterbury (1033 or 1034–1109), born in Italy, scholastic philosopher, entered Benedictine order. Appointed Archbishop of Canterbury in 1093.

Guilbert of Poitiers or Gilbert de La Porrée (1076–1154), French scholastic theologian and bishop of Poitiers who was influential in introducing Aristotelian philosophy in France.

John Duns Scotus (1266–1308), Scottish scholastic theologian, entered Franciscan order. Founder of scholastic system called Scotism, argued that faith is not speculative but an act of will.

William of Auvergne (1180–1249), French philosopher and theologian, bishop of Paris, defended mendicant orders and supported clerical reform. Drew from Aristotle and Avicenna and from Neoplatonism those elements consistent with orthodox Christianity and constructed a theology based on God as essential being.

William of St. Thierry (1085–1148), French theologian and mystic. Entered Benedictine order. Author of works on spirituality and contemplative life, many of them attempting to synthesize Eastern and Western theology and mysticism.

[3]Paul Dirac, *Südkurier*, June 30, 1971.

[4]K. Illies, *VDI-Nachrichten*, July 9, 1971.

[5]Albert Betz quoted by K. Illies in *VDI Nachrichten*, July 9, 1971.

Water: Destiny of the Human Race

[1]*Stuttgarter Zeitung*, June 1973.

[2]*Stuttgarter Zeitung*, March 29, 1973.

[3]*Stuttgarter Zeitung*, January 1, 1973.

[4]*Stuttgarter Zeitung*, March 6, 1972.

[5]*VDI-Nachrichten*, July 18, 1973.

[6]*Stuttgarter Zeitung*, November 10, 1971.

[7]Hermann Ludwig Ferdinand von Helmholtz (1821–1894), German physicist, anatomist, and physiologist. One of the founders of the principle of conservation of energy.

Gustav Ludwig Hertz (1887–1975), German physicist. Was awarded Nobel prize for physics in 1925.

Charles Robert Darwin (1809–1882), English naturalist. Studied animal species and fossils and wrote *On the Origin of the Species*.

Sigmund Freud (1856–1939), Austrian neurologist and psychiatrist. Founder of psychoanalysis.

[8]Auguste Henri Forel (1848–1931), Swiss psychiatrist and entomologist. Studied brain anatomy. Bull. Soc. Vaud. Sci. Nat. *12*/1873 to *40*/1904.

[9]Georg von Békésy (1899–1972), Hungarian-born American physiologist. Helped find treatments for various forms of deafness. Was awarded Nobel prize for physiology in 1961.

[10]Paul Raethjen, *Dynamics of Cyclones*, (Leipzig, 1953). See also "What Is 'Living Water'?" pp. 1–13 in this volume.

[11]August Schmauss, "Wiederkehrende Wetterwendepunkte," *Forschungen und Fortschritte*, 16 (1940), no. 15.

[12]Schmauss, "Wiederkehrende Wetterwendepunkte."

[13]Schmauss, "Biologische Gedanken in der Meteorologie," *Forschungen und Fortschritte*, 21 (1945), no. 1–6.

[14]Norns: goddesses of fate in Norse mythology.

[15]*Song of the Nibelungs*, Canto I, Song 10, trsl. Christy Barnes, *Journal for Anthroposophy*, 30, autumn 1979.

The Warmth Organism of the Earth

[1]Adiabatic changes are temperature changes that take place without addition or loss of heat to or from the surroundings.

[2]Cf. K. Stumpff, A. Schmauss.

[3]Guenther Wachsmuth, *Etheric Formative Forces in Cosmos, Earth and Man: The Path of Investigation into the World of the Living*, vol. I, 2nd ed., (New York: Anthroposophic Press, 1932). We refer the reader here especially to Guenther Wachsmuth's complete description of "the breathing of the earth."

[4]Paul Raethjen, *Dynamics of Cyclones*, (Leipzig, 1953).

[5]August Schmauss, "Wiederkehrende Wetterwende-

punkte," *Forschungen und Fortschritte,* 16 (1940), no. 15, and "Biologische Gedanken in der Meteorologie," *Forschungen und Fortschritte* 21 (1945), no. 1–6. See also "Water: Destiny of the Human Race" pp. 39–63 in this volume.

[6]Schmauss, "Wiederkehrende Wetterwendepunkte."

[7]Johann Wolfgang von Goethe (1749–1832), German poet. His most famous work is the verse drama *Faust*. Goethe also wrote extensively on botany, geology, optics, and other scientific topics.

[8]L. J. Henderson, *Die Umwelt des Lebens,* (Wiesbaden, 1914).

[9]Hagen, "Naturwissenschaftliche Grenzen des Wachstums," *VDI-Nachrichten,* April 19, 1974.

[10]P. W. Pohl, *Wärmelehre,* (Göttingen, 1946).

[11]R. Riedl, "Energie, Information und Negentropie in der Biosphäre," *Naturwissenschaftliche Rundschau,* 10, (Stuttgart, 1973).

[12]Riedl, "Energie, ..."

[13]Rudolf Steiner, *Warmth Course,* No 321 in the Collected Works, (Spring Valley: Mercury Press, 1988).

[14]Steiner, *Warmth Course.*

[15]Friedrich Schiller, (1759–1805), German poet, playwright, and critic. The poem in the text is entitled *"Das Höchste"* ("The Highest").

[16]Hagen, "Naturwissenschaftliche Grenzen des Wachstums."

[17]Riedl, "Energie, ..."

[18]Rudolf Steiner, *The Bridge between Universal Spirituality and the Physical Constitution of Man,* Lecture 2, No. 202 in the Collected Works, repr. (Spring Valley: Anthroposophic Press, 1979).

[19]Riedl, "Energie, ..."

Keeping the Earth the Place of Life

[1]Friedrich Schiller, *On the Aesthetic Education of Man,* (Oxford: Oxford Univ. Press, 1967).

[2]Nigel Calder, *The Weather Machine,* (New York: Viking, 1974).

[3]Calder, p. 27.
[4]Julius Robert Mayer (1814–1878), German physician and physicist. His work was the foundation for the discovery of the first law of thermodynamics.
[5]Calder, p. 64.
[6]Calder, p. 67.
[7]Calder, p. 124.
[8]L. J. Henderson, *Die Umwelt des Lebens,* (Wiesbaden, 1914).
[9]Rudolf Steiner, *Entsprechungen zwischen Mikrokosmos und Makrokosmos,* 16 Lectures given in Dornach, No. 201 in the Collected Works, (Dornach, Switzerland: Rudolf Steiner Verlag, 1987). Not yet translated.
[10]Calder, p. 138.

The Human Role in the Earth's New Connection with the Cosmos

[1]Johann Wolfgang von Goethe (1749–1832), German poet and playwright. Visited Italy 1786–1788 and again in 1790 and was inspired by the beauty of its countryside and historical sites. Published his journal of the voyage. Goethe also wrote extensively on botany, geology, optics, and other scientific topics.
[2]Georg Kleemann in *Stuttgarter Zeitung,* March 17, 1977.
[3]Rudolf Steiner, *Truth and Knowledge,* 2nd ed., (Blauvelt, N.Y.: Steinerbooks, 1981) and *Philosophy of Spiritual Activity,* (Hudson, N.Y.: Anthroposophic Press, 1986).
[4]Rudolf Steiner, *A Theory of Knowledge Implicit in Goethe's World Conception,* 3rd ed., (Spring Valley, N.Y.: Anthroposophic Press, 1978), *Die Evolution vom Gesichtspunkte des Wahrhaftigen,* 5 lectures, Berlin 1911, No. 132 in the Collected Works, (Dornach, Switzerland: Rudolf Steiner Verlag, 1987); not yet translated. *Knowledge of the Higher Worlds,* repr., (Hudson, N.Y.: Anthroposophic Press, 1986).
[5]Theodor Schwenk, "What Is 'Living Water'?" pp. 1–13 in this volume.

Motion Research: Its Course and Aims over Twenty Years

[1]*Stuttgarter Zeitung*, February 1, 1978.

[2]*Stuttgarter Zeitung*, February 1, 1978.

[3]*Frankfurter Allgemeine Zeitung*, January 22, 1979.

[4]Theodor Schwenk, "Keeping the Earth the Place of Life" pp. 97–120 in this volume.

[5]*Welt am Sonntag*, February 18, 1979.

[6]K. Illies in *VDI-Nachrichten*, July 9, 1971.

[7]Theodor Schwenk, *The Basis of Potentization Research*, (Spring Valley, N.Y.: Mercury Press, 1988).

[8]Theodor Schwenk, *The Basis of Potentization Research*, (Spring Valley, N.Y.: Mercury Press, 1988). George Adams and Olive Whicher, *The Plant Between Sun and Earth*, 2nd ed., (London: Rudolf Steiner Press, 1980). Theodor Schwenk, *Sensitive Chaos*, (London: Rudolf Steiner Press, 1976).

[9]Johannes Kepler (1571–1630), German astronomer who discovered Kepler's laws of planetary motion. Postulated ray theory of light to explain vision.

[10]Rudolf Steiner, *The Etherisation of the Blood: The Entry of the Etheric Christ into the Evolution of the Earth*, repr., (London: Rudolf Steiner Press, 1985).

[11]Rudolf Steiner, *Entsprechungen zwischen Mikrokosmos und Makrokosmos*, 16 Lectures given in Dornach, no. 201 in the Collected Works, (Dornach, Switzerland: Rudolf Steiner Verlag, 1987). Not yet translated.

[12]"For the earnest expectation of the creature waiteth for the manifestation of the sons of God, for we know that the whole creation groaneth and travaileth in pain together until now." St. Paul.

Water as the Element of Life

[1]For further details, see Theodor Schwenk, *Sensitive Chaos*, (London: Rudolf Steiner Press, 1976).

[2]See Theodor Schwenk, *Bewegungsformen des Wassers:*

Nachweis feiner Qualitätsunterschiede mit der Tropfenbildmethode, (Stuttgart: Verlag Freies Geistesleben, 1967), not yet translated.

Water Sustains All

[1]Thales of Miletus (625–547 B.C.), Greek philosopher and scientist. He taught that water, or moisture, was the one element out of which the world was formed.

[2]*Faust,* drama in verse by the German poet Johann Wolfgang von Goethe (1749–1832). It is based on the legend of Dr. Faustus, a magician said to have performed miracles with the help of the devil.

[3]O. J. Hartmann, "Das Wasser: Träger des Lebens," *Die Kommenden* 30 (Heft 5) 1976: 19–21.

[4]K. Lang, *Wasser, Mineralstoffe, Spurenelemente,* (Darmstadt: UTB, 1974, vol. 341).

[5]see Lang, *Wasser, ...*

[6]see Lang, *Wasser, ...*

[7]Theodor Schwenk, "What Is 'Living Water'?" pp. 1–13 in this volume.

[8]Theodor Schwenk, "The Warmth Organism of the Earth," pp. 65–96 in this volume.

[9]M. Debus, "Das Wasser," *Archiv Badewesen,* 24 (1971):721–727.

[10]W. A. P. Luck, ed., "Structure of Water and Aqueous Solutions," *Proceedings Intern. Sympos.,* Marburg, July 1973. (Weinheim: 1974).

[11]see Theodor Schwenk, "What Is 'Living Water'?" pp. 1–13 in this volume.

[12]Theodor Schwenk, *Sensitive Chaos,* (London: Rudolf Steiner Press, 1976). The following statements are documented by 30 photographs, most of which are taken from the book referred to here.

[13]Theodor Schwenk, *Bewegungsformen des Wassers: Nachweis feiner Qualitätsunterschiede mit der Tropfenbildmethode,* (Stuttgart: Verlag Freies Geistesleben, 1967).

[14]see Theodor Schwenk, "What Is 'Living Water'?" pp. 1–13 in this volume.

[15]Anonymous (Fr. Lbg.), "Zwischen Schwerkraft und kosmischen Strahlen," *Südkurier*, April 23, 1976.
[16]Th. Schwenk et al., publication in preparation.
[17]Th. Schwenk, *Bewegungsformen des Wassers;* Th. Schwenk, publication in preparation; Th. Schwenk, "Hat die Erforschung der Konstellationswirkungen im Irdischen praktische Bedeutung?" *Sternkalender 1976/77*, (Dornach 1975), pp. 83–87. See also G. Unger, "Konstellationsforschung," *Sternkalender 1970/71*, (Dornach, 1969), pp. 83–87.
[18]Rudolf Steiner, Collected Works (numerous titles, many not yet published or translated) (Dornach, Switzerland: Rudolf Steiner Verlag).
[19]Th. Schwenk in a conversation.
[20]see note 13 and the following articles in this book.

Testing for Water Quality: The Drop-Picture Method

[1]Theodor Schwenk, *Bewegungsformen des Wassers: Nachweis feiner Qualitätsunterschiede mit der Tropfenbildmethode*, (Stuttgart: Verlag Freies Geistesleben, 1967), not yet translated; "Neue Qualitätsprüfung von Trinkwasser," *VDI-Nachrichten* 23, No. 15, (Düsseldorf, 1969), p. 12.
[2]Wolfram Schwenk, "Studying the Behavior of Water," pp. 205–213 in this volume.
[3]U. Hässelbarth, "Überlegungen zur Definition des Trinkwassers," Wasserfachliche Aussprachetagung Basel 1975, DVGW-Schriftenreihe *Wasser*, 10, (Eschborn, 1976), pp.83–87.
[4]*Deutsche Normen*, "DIN 2000, Zentrale Wasserversorgung: Leitsätze für Anforderungen an Trinkwasser. Planung. Bau und Betrieb der Anlagen," (Berlin: Verlag Beuth Vertrieb, 1973).
[5]F. Morell, "Qualitätsbestimmung von Wasser durch die VINCENT-Methode," talk at the Hessische Winterseminar, October 27, 1979.
[6]Theodor Schwenk, *Jahresbericht des Instituts für Strömungs-*

wissenschaften 1970, Herrischried, Manuskriptdruck; M. Schütz and P. Schütz, *Experimentelle Studien mit der Tropfenbildmethode im Zusammenhang mit einer Sonnen- und Mondfinsternis,* Forschungs- und Versuchsanstalt der Stadt Wien, 1977; G. Unger, "Konstellationsforschung," *Sternkalender 1970/71,* (Dornach, Switzerland: Philosophisch-Anthroposophischer Verlag, 1969), pp. 83–87. See also Theodor Schwenk, "Hat die Erforschung der Konstellationswirkungen im Irdischen praktische Bedeutung?" *Sternkalender 1976/77,* (Dornach, Switzerland: Philosophisch-Anthroposophischer Verlag, 1975), pp. 83–87.

[7]Theodor Schwenk, *Sensitive Chaos,* (London: Rudolf Steiner Press, 1976), and see also his lectures and talks at the Institute for Flow Research in Herrischried 1967–1979. Wolfram Schwenk, "Water Sustains All," pp. 177–191 in this volume.

Studying the Behavior of Water

[1]Theodor Schwenk, *Bewegungsformen des Wassers: Nachweis feiner Qualitätsunterschiede mit der Tropfenbildmethode,* (Stuttgart: Verlag Freies Geistesleben, 1967), not yet translated; "Neue Qualitätsprüfung von Trinkwasser," *VDI-Nachrichten* 23, No. 15, (Düsseldorf, 1969), p. 12. See also D. Rapp and P. E. M. Schneider, *Das Tropfenbild als Ausdruck harmonischer Strömungen in dünnen Schichten,* Max-Planck-Institut für Strömungsforschung Göttingen, report no. 102/1974.

[2]H. J. Smith, *A Study of some of the Parameters Involved in the Drop-Picture Method,* Max-Planck-Institut für Strömungsforschung Göttingen, report no. 111/1974. H. Smith, *The Hydrodynamic and Physico-Chemical Basis of the Drop Picture Method,* Max-Planck-Institut für Strömungsforschung Göttingen, report no. 8/1975.

[3]Deutscher Normenausschuss, "Zentrale Trinkwasserversorgung: Leitsätze für Anforderungen an Trinkwasser, Planung, Bau und Betrieb der Anlagen," *Deutsche Normen, DIN 2000,* (Berlin: Verlag Beuth Vertrieb, 1973).

[4]W. Drobek, "Gedanken über eine Grosstadt-Wasserversorgung um die Jahrtausendwende," *Das Gas- und Wasserfach*, 108 (1967) and 109 (1968).

[5]Wolfram Schwenk, *Vergleichende Gewässeruntersuchungen mit der Tropfenbildmethode und hydrobiologisch-stoffwechseldynamischen Verfahren*, publication in preparation.

[6]see note 6 of "Testing for Water Quality: The Drop-Picture Method" and W. Schwenk, "Water Sustains All", pp. 177–191 in this volume.

Water as a Nutrient

[1]R. D. Rurainski et al., "Über das Vorkommen von natürlichen und synthetischen Östrogenen im Trinkwasser," *Das Gas- und Wasserfach: Wasser/Abwasser* 118, no. 6/1977, pp. 281–291.

[2]S. Feldhoff and K. D. Koss, "Zur Problematik des Nitrats im Trinkwasser," *Forum Städte Hygiene* 33, (Hannover, 1982), pp. 96–102.

[3]H. Althaus, "Hygiene der Trinkwasserversorgung," *Das Gas- und Wasserfach: Wasser/Abwasser* 119, no. 11/1978, pp. 542–547.

[4]Rat von Sachverständigen für Umweltfragen beim Bundesministerium des Innern, Bonn, "Umweltprobleme der Nordsee," summary of the report, *Umweltbrief* 22/1980.

[5]Bundesministerium des Innern, *Umwelt*, 58/1977, 16, Bonn.

[6]J. Zobrist and W. Stumm, "Wie sauber ist das Schweizer Regenwasser?" *Neue Zürcher Zeitung*, 146, June 27, 1979.

[7]U. S. Council on Environmental Quality, *Global 2000: Report to the President*, (Washington, D.C.: Government Printing Office, 1980–1981).

[8]G. Picht, "Industrieplanung in der ökologischen Krise," *Umwelt*, 1/1972, pp. 28–31.

[9]J. Illies, *Umwelt und Innenwelt*, (Freiburg: Herder-Verlag, 1974).

[10]Rudolf Steiner, *Towards Social Renewal: Basic Issues of the Social Question*, No. 23 in the Collected Works, (London:

Rudolf Steiner Press, 1977); *The Renewal of the Social Organism*, No. 24 in the Collected Works, (Spring Valley, N.Y.: Anthroposophic Press, 1985).

[11]Rudolf Steiner, *Agriculture: A Course of Eight Lectures*, No. 327 in the Collected Works, 3rd ed., repr., (London: Biodynamic Agricultural Association, 1984).

[12]Bundesverband der deutschen Gas- und Wasserwirtschaft, *Trinkwasser*, Werbefaltblatt, Bonn.

[13]K. Haberer, "Neuere technologische Entwicklungen in der Wasseraufbereitung," *Das Gas- und Wasserfach: Wasser/Abwasser* 120, no. 3/1979, pp. 103–113.

[14]Bundesministerium für Jugend, Familie und Gesundheit, "Verordnung über Trinkwasser und über Brauchwasser für Lebensmittelbetriebe (Trinkwasser-Verordnung) vom 31.1.-1975," *Bundesgesetzblatt I*, p. 453; most recently changed through Article 1 of *Verordnung* of June 25, 1980, *Bundesgesetzblatt I*, p. 764. "Die Trinkwasserverordnung: Erfahrungsberichte verschiedener Autoren," *Forum Städte Hygiene* 33, no. 2/1982.

[15]Verein Deutscher Ingenieure, "10. Jahrestagung über Umweltanalytik," Dortmund 1980, *VDI-Nachrichten* 25/1980.

[16]Deutscher Normenausschuss, "Zentrale Trinkwasserversorgung: Leitsätze für Anforderungen an Trinkwasser, Planung, Bau und Betrieb der Anlagen," *Deutsche Normen, DIN 2000*, (Berlin: Verlag Beuth Vertrieb, 1973).

[17]Translator's note: The German word for food or groceries, "*Lebensmittel*" literally means "life-conveying" or "life-mediating."

[18]U. Hässelbarth, "Überlegungen zur Definition des Trinkwassers," Wasserfachliche Aussprachetagung Basel 1975, DVGW-Schriftenreihe *Wasser*, 10, (Eschborn, 1976), pp.83–87.

[19]Theodor Schwenk, "Neue Qualitätsprüfung von Trinkwasser," *VDI-Nachrichten* 23, No. 15, (Düsseldorf, 1969), p. 12; Wolfram Schwenk, "Studying the Behavior of Water," pp. 205–213 in this volume.

[20]Rudolf Steiner, *Warmth Course*, No. 321 in the Collected Works, (Spring Valley: Mercury Press, 1988); Rudolf Steiner, *Die Weihnachtstagung zur Begründung der Allgemeinen Anthro-*

posophischen Gesellschaft 1923/24, No. 260 in the Collected Works, (Dornach, Switzerland: Rudolf Steiner Verlag, 1987), not yet translated. Rudolf Steiner, *Knowledge of the Higher Worlds and Its Attainment,* No. 10 in the Collected Works, (Spring Valley, N.Y.: Anthroposophic Press, 1983).

[21]Rudolf Steiner, *Die Verantwortung des Menschen für die Weltentwickelung,* No. 203 in the Collected Works, (Dornach: Rudolf Steiner Verlag, 1982), not yet translated.

[22]F. Hofmann, "Goethe als Naturforscher," *Badische Zeitung,* Freiburg, March 20/21, 1982.

[23]Theodor Schwenk, *Bewegungsformen des Wassers: Nachweis feiner Qualitätsunterschiede mit der Tropfenbildmethode,* (Stuttgart: Verlag Freies Geistesleben, 1967).

[24]Theodor Schwenk, *Bewegungsformen des Wassers: Nachweis feiner Qualitätsunterschiede mit der Tropfenbildmethode,* (Stuttgart: Verlag Freies Geistesleben, 1967). P. E. M. Schneider, *Sechs Instabilitätsphasen eines Ringwirbels als Grundlage für eine Klassifikation der Schwenkschen Tropfenbilder,* Max-Planck-Institut für Strömungsforschung Göttingen, Report no. 9/1976.

[25]R. Rautenstrauch, "Beobachtungen an Tropfenbildern von Pflanzensäften," *Elemente der Naturwissenschaft* 25, (Dornach, 1976), pp. 35–42; "Über Formtendenzen in Tropfenbildern," *Elemente der Naturwissenschaft* 31, (Dornach, 1979), pp. 10–23.

[26]Theodor Schwenk, *Bewegungsformen des Wassers: Nachweis feiner Qualitätsunterschiede mit der Tropfenbildmethode,* (Stuttgart: Verlag Freies Geistesleben, 1967); "Neue Qualitätsprüfung von Trinkwasser," *VDI-Nachrichten* 23, No. 15, (Düsseldorf, 1969), p. 12. Wolfram Schwenk, "Testing for Water Quality: The Drop-Picture Method," pp. 193–203 in this volume. U. Hoesle, *Die Tropfenbildmethode im Vergleich zu anderen Testmethoden zur Qualitätsbeurteilung von Abwasser und ihre Anwendung im Klärwerk,* Master's Thesis, University at Hohenheim, summer 1977.

[27]D. Rapp and P. E. M. Schneider, *Das Tropfenbild als Ausdruck harmonischer Strömungen in dünnen Schichten,* Max-Planck-Institut für Strömungforschung Göttingen, report no. 102/1974. H. J. Smith, *A Study of some of the Parameters Involved*

in the Drop Picture Method, Max-Planck-Institut für Strömungsforschung Göttingen, report no. 111/1974; H. J. Smith, *The Hydrodynamic and Physico-chemical Basis of the Drop-Picture Method,* Max-Planck-Institut für Strömungsforschung Göttingen, report no. 8/1975.

[28]Wolfram Schwenk, "Studying the Behavior of Water," pp. 205–213 in this volume and Wolfram Schwenk, "Testing for Water Quality: The Drop-Picture Method," pp.193–203 in this volume. Wolfram Schwenk, *Jahresbericht für 1977 des Instituts für Strömungswissenschaften,* Herrischried.

[29]Wolfram Schwenk, "Von der Bedeutung des Begriffs des Sensiblen Chaos für die heutige Naturforschung," series of talks given in summer 1981 in Herrischried, publication pending.

[30]H. Haken, *Synergetik,* (Berlin: Springer Verlag, 1982); "Synergetik, a New Science," *Stuttgarter Zeitung,* October 10, 1979.

[31]George Adams, *Physical and Ethereal Spaces,* (London: Rudolf Steiner Press, 1965); Theodor Schwenk, *Sensitive Chaos,* (London: Rudolf Steiner Press, 1976); Theodor Schwenk, *The Basis of Potentization Research,* (Spring Valley, N.Y.: Mercury Press, 1988).

[32]Theodor Schwenk, *Sensitive Chaos,* (London: Rudolf Steiner Press, 1976).

[33]Rudolf Steiner, *Man as Symphony of the Creative Word,* No. 230 in the Collected Works, (London: Rudolf Steiner Press, 1970).

[34]Rudolf Steiner, *Warmth Course,* No. 321 in the Collected Works, (Spring Valley: Mercury Press, 1988); Rudolf Steiner, *Agriculture: A Course of Eight Lectures,* No. 327 in the Collected Works, 3rd ed., repr., (London: Bio-Dynamic Agricultural Association, 1984).

[35]L. Kolisko, "Workings of the Stars in Earthly Substances," *Experimental Studies from the Biological Institute of the Goetheanum,* (Stuttgart: Orient-Occident Verlag, 1928).

[36]Theodor Schwenk, *The Basis of Potentization Research,* (Spring Valley, N.Y.: Mercury Press, 1988); Theodor Schwenk, *Bewegungsformen des Wassers: Nachweis feiner*

Qualitätsunterschiede mit der Tropfenbildmethode, (Stuttgart: Verlag Freies Geistesleben, 1967). Theodor Schwenk, *Jahresbericht des Instituts für Strömungswissenschaften 1970,* Herrischried, Manuskriptdruck. Theodor Schwenk, "Hat die Erforschung der Konstellationswirkungen im Irdischen praktische Bedeutung?" *Sternkalender 1976/77,* (Dornach, Switzerland: Philosophisch-Anthroposophischer Verlag, 1975), pp. 83–87.